The ENDURING FORESTS

The ENDURING FORESTS

Northern California, Oregon, Washington, British Columbia, and Southeast Alaska

RUTH KIRK ▪ EDITOR CHARLES MAUZY ▪ PHOTO EDITOR
FOREWORD BY ROBERT MICHAEL PYLE

THE MOUNTAINEERS & THE MOUNTAINEERS FOUNDATION

Published by
The Mountaineers
1001 SW Klickitat Way
Seattle, Washington 98134

0 9 8 7 6
5 4 3 2 1

Published simultaneously in Canada by Douglas & McIntyre, Ltd., 1615 Venables Street, Vancouver, B.C. V5L 2H1

Published simultaneously in Great Britain by Cordee, 3a DeMontfort Street, Leicester, England, LE1 7HD

Manufactured in Hong Kong by Bookbuilders Ltd.

Edited by Cynthia Newman Bohn
Maps by Michelle Taverniti
Cover design by Elizabeth Watson
Book design by Alice C. Merrill
Typography by The Mountaineers Books

Cover photograph: *Old-growth forest, Soleduck Valley, Olympic Peninsula* (Photo by Charles Mauzy)
Frontispiece: *Side channel of Quinault River, Olympic Peninsula* (Photo by Ruth Kirk)

Type set in Minion and Poetica

Library of Congress Cataloging-in-Publication Data

The enduring forests : Northern California, Oregon, Washington, British Columbia, and Southeast Alaska / Ruth Kirk, editor ; Charles Mauzy, photo editor.
p. cm.
Includes bibliographical references (p.) and index.
ISBN 0-89886-467-4
1. Forest ecology—Northwest Coast of North America. 2. Forests and forestry—Northwest Coast of North America. 3. Old growth forests—Northwest Coast of North America. I. Kirk, Ruth.
QH104.5.N6E54 1996
547.5'2642'09795—dc20
95-49553
CIP

Spring alder tassels

Printed on recycled paper

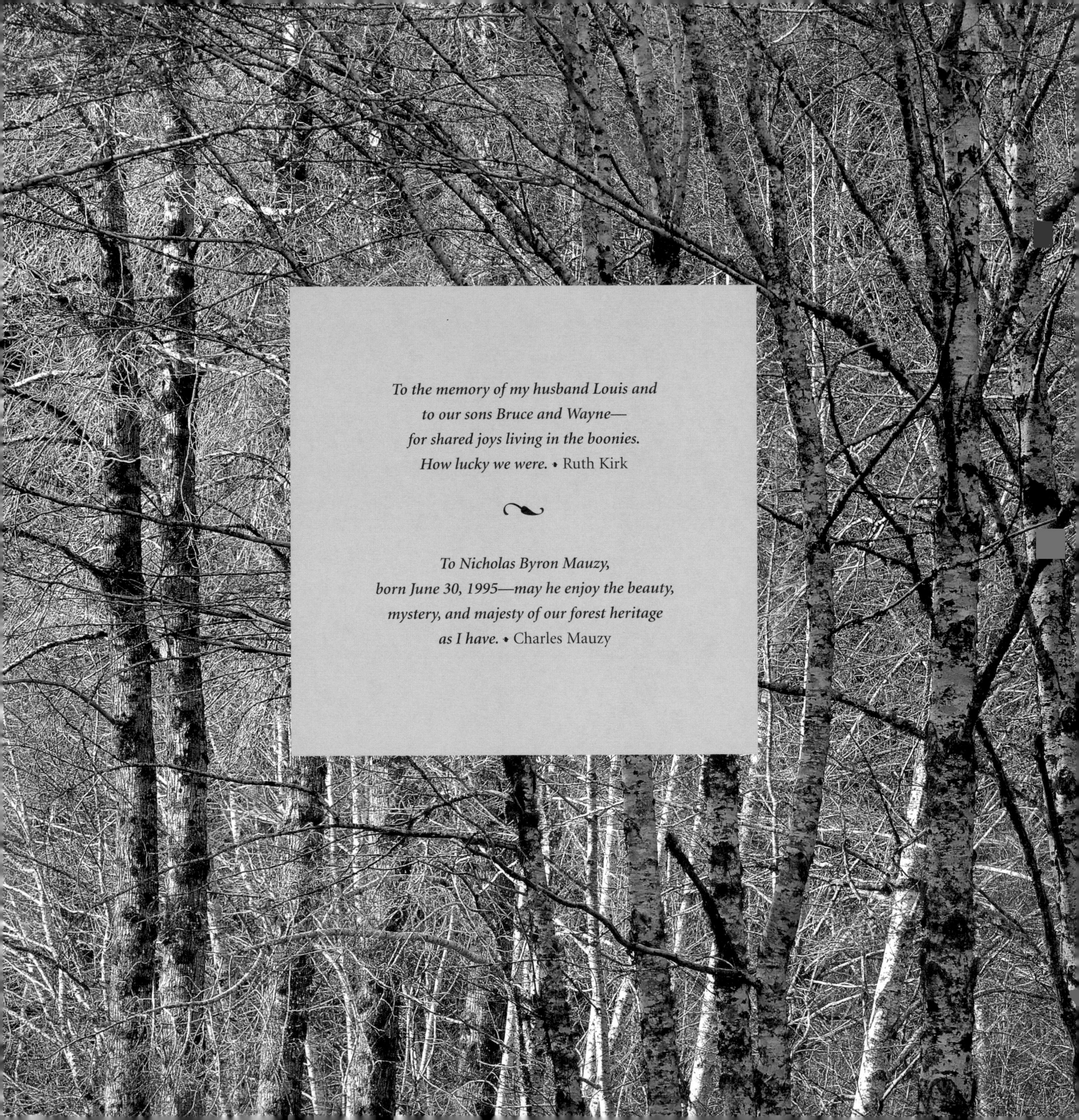

To the memory of my husband Louis and
to our sons Bruce and Wayne—
for shared joys living in the boonies.
How lucky we were. • Ruth Kirk

To Nicholas Byron Mauzy,
born June 30, 1995—may he enjoy the beauty,
mystery, and majesty of our forest heritage
as I have. • Charles Mauzy

Photography Credits

© as follows:

Gary Braasch: pages 58 (top right and lower left), 64, 106–107, and 127

Carr Clifton: pages 14, 142, and 158

Daniel Cox/Natural Selection: page 149

Adrian Dorst: page 132

Adrian Dorst/LightHawk: page 138

Jerry F. Franklin: pages 89, 97, 116, 123, and 131

Dennis Frates/Natural Selection: page 8

David J. Job: pages 140–141

Johnny Johnson/Natural Selection: page 157

Ruth Kirk: pages 35, 80, 82, 104, 112, 115, 124, 137, 151, 152, 162, 165, and frontis.

Charles Mauzy: pages 5, 20–21, 30, 33, 40, 57, 74–75, 81, 85, 86, 92, 93, 99, 170, and cover

Joe McDonald/Natural Selection: page 90

David Muench 1995: pages 22, 27, 36, 39, 42–43, 44, 50, 60, 63, 70, 76, 108, 120, 161, and 164

Pat O'Hara: pages 7 and 103

Trygve Steen: pages 53, 54, and 69

Greg Vaughn: page 49

Kennan Ward: page 167

Art Wolfe: page 58 (upper left)

Summer spruce needles

CONTENTS

I came up here to sit by myself and I saw the logger-men's tag on a little tree, that they were going to put a road through here. So I met with the council and I told them, "This is the last one. Not one more step backward." And the chiefs stood up and they said, "This is where we'll take our stand."

I had a dream. My hand was reaching out and there were hands coming down, but they never touched. Then after the Victoria conference (a meeting of Haisla leaders, forest scientists, and provincial and timber company representatives) I shook hands and they all touched this brown hand, those scientists, and I knew my dream had come true.
I believe in dreams.

—Chief Cecil Paul, born at Kitlope,
Northern British Columbia, speaking in 1993

Foreword

In the clamor and clangor over the future of the remaining old-growth forests, we tend to forget the brutal fact that most of the land we live with has already been used up. Not completely exhausted, for land seldom reaches such a state. *Something* will grow on the most abused soils, and that will compost itself for the next level of complexity. In this way, had we time to wait, we might see our cutover estate reclaim the look of mature woods, even the demeanor we ennoble with the designation "old growth." The species complement would be different, for extinctions would have occurred, and other details would differ from the original—but the ultimate outcome would be hoary and sylvan: the Big Bush would be back.

But we haven't got the time. Unlike that of trees, life is short for us. Under the best circumstances, succession takes a long duration to build a real forest. If you lived where I do, you could be forgiven for imagining otherwise. Here in southwest Washington's Willapa Hills, a modest flange of the Coast Ranges, the old growth has long since fled. Mostly privately owned and managed, these deep green creases and mountainettes were scoured for their dense and massive Douglas-firs, western hemlocks, western redcedars, and Sitka spruces before the northern spotted owl ever thought of alighting in the public mind. Yet beneath the balm and burden of four meters of annual rainfall, regrowth explodes. On my own old Swedish farmstead, the river terrace left alone for a season or two springs into instant woodland as native alders and English oaks struggle with hemlocks and spruces for space and sunbeams. The practice many Willapa families adopt, of blasting back all wild greens and replacing them with massive lawns, is merely an attempt to hold the growth at bay. "It'll grow ya right out," I've been told by a logger who wields a chain saw all day, a lawnmower and trimmers when he gets home.

On the managed forestland too, regrowth comes rapidly, except where the

Autumn maple leaves

headwalls of valleys have failed in erosive seizures brought on by saturation and steep-slope logging. The speedy greening of the cutover lands almost makes you think that it will all come back. But look closely. Walk into these doghair plantations, and you will see the problem with easy assumptions of sustainability: these are not forests, after all. They haven't any of the complexity, richness, subtlety, or whimsy of the greenwood as we knew it. The industrial woods may be deep, they may be dark, but they are not diverse. You couldn't imagine Sasquatch lurking in such a stand; your children would not look for leprechauns behind the shamrock sorrel. It takes time, lots of it, to make such a place. No one has time to wait for forests any more.

In this remarkable book, from the testimony of writers and photographers who know, we come face to face with the stuff of real forests. We peer into the joineries that link dominant species and their soils into those looming entities where towering transpiration takes place. We follow scientists who go up in steel cranes and polypropylene cradles to find out who lives where and how the whole improbable thing works. We eavesdrop on fire and whisper after salmon, we are arrested with the faithful who stand beside the First Nation peoples to protect the First Trees. And we ponder what it means to be dependent on forests—for life, for livelihood, for the glory and fascination of the world, for the paper pulp on which to print our pleas for old-growth protection. In the end, through the articulate voices of poets and ecologists and administrators and activists, we come to a deeper sense of what it means to belong to Robin Goodfellow's band: the growing company of people, merry men and others, who will not stand for the dismantling of the greenwood. More and more, the reformers are coming to include not only outlaw demonstrators and maverick ecologists but also enlightened foresters, managers, and landowners themselves. As we learn what the word "forest" really means, we begin to change how we comport ourselves in the woods. And that is encouraging.

If President Clinton's Forest Plan should survive a Congress increasingly eager to dismantle national parks and penetrate the Arctic National Wildlife Refuge, a substantial measure of the remaining old woods will be protected. This is thanks to the northern spotted owl, since that is the way the Endangered Species Act—itself imperiled—works. Of course, the forests are no more reserved "for the owl" than for *Hemphillea,* the uncommon shell-bearing slug that occupies the duff of old true-fir forests; than for martens and cougars and tree voles; than for that response of ours we call spiritual. But United States law says we must identify individual species at risk, then define critical habitat necessary for their recovery; Canadian law has no comparable provision. Nowadays—on both sides of the border—people are beginning to talk about ecosystem management, as if we knew what *that* meant. One thing seems sure: a system-wide approach might avoid staking whole forests' futures on the occurrence of a single kind of creature; but the job it requires is tougher yet to accomplish or monitor, and to do it right would mean much *more* land reserved from the saw, not less.

In our rush to define and conserve systems, we must not demean the importance of knowing exactly *who* lives in the bush. For many years I have been involved in ongoing studies of the biogeography of Northwest invertebrates. Even the butterflies, arguably the best-known group of insects, leave major gaps in our knowledge of their distribution and ecology. Moths, much more so. Certain other groups are nearly unknown. While we will never draft management plans for all of the invertebrates in a working woodland, how can we begin to think we know its internal workings as we stand by so vastly uninformed, blissed out in our ignorance? University and government efforts at biological survey have withered in recent years, in favor of sexier science. Yet recent studies give an idea of the job ahead. In the upper Carmanah Valley of British Columbia—these pages tell us—Neville Winchester, of the University of Victoria, collected more than a million arthropods from five big spruces, bridged by canopy platforms, and the forest floor nearby. Of these, 67 have been confirmed as new species, with ten times as many expected. Such is our knowledge of what E. O. Wilson calls "the little things that run the world!"

From my vantage in the logged-over, brushed-up land of Willapa, I place the sack of the woods at the feet of three sad facts. The first is the very weight of our own grievous numbers, pressing down on every other species on the globe, making unreasonable demands on the fiber that also happens to bond the world's major bank of nitrogen and carbon. The second is a species of greed and short-sightedness that is willing to convert Tongass giants into bolts of rayon, at our loss and someone's great gain; willing even to mine the boreal smallwood—the aspen, birch, and taiga conifers of the northland—to feed our unquenchable appetite for pulp. And the third would be our dissociation from the fabric of the forest itself. As Ruth Kirk has asked me, "How can we wisely decide policy in a democracy where most people no longer know the earth personally? Outdoor 'experience' is primarily found on television rather than by sweating, swatting deer flies, and reveling in natural beauty." She has it exactly right. How can a child who has never known a wren grow up caring about a condor? When memories of the deep, dark forest come from CD-ROMs instead of storybooks and walks with elders, how can children be expected to mourn the passing of the real thing?

What makes this particular storybook important to me is its power to convey us through the spruce portals, down the moss trail, into the heart of the actual. For no one, having read these tales of real-life diversity, having swum with their eyes in the green pools of color and texture depicted by the plates, will be able to resist going forth to experience the kinds of woods from which such words and pictures flow. We live in a time and place where the virtual supersedes the actual, and we are being asked to accept it. For all their value, television shows about nature are mere projections. Interactive computer tools might be entertaining and novel, but what they interact with is not *there.* And the new tree-cover now blanketing many western landscapes consists merely of virtual forests. Doomed by genetics, soil-scorching slash burns, herbicides, and cutting cycles that last a mere quarter-century at the low end and seldom exceed twice that as sawlogs give way to pulp and chipboard, these plantations will never become actual forests as we once knew them: forests such as this book portrays.

You will not find the bruised but green-bandaged Willapa Hills in these pages. As a post-old-growth landscape, the Willapa represents what we hope the enduring forests will never become. In fact, we have managed to save a few lingering patches of ancient forest here, but the total acreage barely requires a single comma. Much of the rest was on its way back to maturity when I came here in the late seventies, but the liquidation of the older second-growth in the past decade has been amazing to behold: fifty years of jobs and habitat of growing complexity, used up in no time for maximum short-term profit by the insurance firms and holding companies that bought the stumpage from the old timber outfits. We shall watch our woodland legacy stretch away, into the distance and the future, densely packed little evergreens furring the damaged slopes like stunted taiga, where there should be rain-forest giants.

And so it will be everywhere one day, unless we as a people decide to keep the real woods, and not settle for the virtual . . . as the lovers of the remaining California redwoods, Opal Creek in Oregon's Cascades, the coastal corridors of Washington's Olympic Peninsula, British Columbia's Kitlope and South Moresby, Southeast Alaska's Admiralty Island, have all decided. We know that we are all part of the outcome, for our neighbors' jobs and our own complicity in human affairs cause all of us to draw upon the largesse of the trees. As the blues singer Keb' Mo' aptly puts it, we are all victims of comfort. But if our needs cannot be met on the vast acreage of managed forestlands, then our expectations will have to be adjusted. It is time for us to decide together to treat our woods—the cutover as well as the untouched—as if they were really there. Nothing else makes sense.

For we do not have to settle for old-style clearcuts, old-fashioned dull plantations, or old thinking when it comes to trees. The pace of real forest inquiry has never been greater, nor more timely. In this book you will learn the crystalline sum of decades of old-growth research by Jerry Franklin, who brought the field to respectability. You will climb with Nalini Nadkarni into the rain-forest canopy to learn for the first time how trees and their epiphytes actually interact. And you will confront "New Perspectives" in forestry that are reformulating how we will extract fiber from the forest while leaving a forest behind. The retention of live trees, snags, and woody debris on a logging show, while far from universal, is becoming common enough not to excite attention. The lessons from the old growth are being applied in order to leave timber sales as much like functioning forests as possible. If our appetites for pulp and profit can be subordinated to the needs of the future, and if our resolve can match our new knowledge, even on commercial lands we may yet see something like real forests arising again. From an ecological standpoint, this fresher sort of forestry produces landscapes that are not as good as old growth, but are a lot better than doghair. When I go into such forests, I find hope for diversity.

Let this book, then, with its messages from the real places, take you by the hand into the forest. It is a friendly place after all, not the abode of ogres—they live far outside, in a money-misted land where the forest cannot be seen for the fees. Go afield with these messengers, find your own paths among the ferns and needles. Ask why this tree grows here, that one over there. Contemplate fire scars, meditate on the moss. Then go home, bathed in the terpenes of firs, the oils of redcedar. And determine that in whatever way you can, you will help these forests to endure.

—Robert Michael Pyle
Gray's River, Washington

It is essential that we re-examine ethically what we have inherited, what we are responsible for, and what we will pass on to coming generations. Clearly ours is a pivotal generation.

—The Dalai Lama,
in *My Tibet,* by Galen Rowell,
University of California Press, 1992

Preface

Every book moves from conception through gestation to bound volume in its own way. In this case Northwest photographer Charles Mauzy, who grew up in the redwood country around Crescent City, California, wanted to produce a book of such compellingly beautiful photographs that the public would see the old-growth forests of the West in a new way. The geographic reach of the book was to extend across a wide swath of western North America, and the text was to be something of a battle cry for preservation. The books and conservation divisions of The Mountaineers expressed encouragement, and Ruth Kirk joined as overall editor. She brought to the project a background of having lived in the forests of Mount Rainier and Olympic National Parks in the 1950s and 1960s with her ranger-naturalist husband and having written copiously about the Northwest. Work commenced.

The essence of the original concept remained intact but took on new definition. Lowland forests, rather than those of the montane and subalpine regions, became the book's focus, and its geographic scope narrowed to the region from the redwoods of northern California to the tidewater glaciers and Sitka-spruce forests of Southeast Alaska. This vast Pacific strip is the exclusive range of the spruce as well as the home of other classic Northwest trees, for example, Douglas-fir, western hemlock, and western redcedar. Because the latter species also grow inland and because of the apparently dire future of inland ponderosa pine forests, we decided to include forests east of the Coast and Cascade mountains of Oregon and Washington and the Coast Mountains of British Columbia.

As conception segued into action, the approach shifted away from delineating specific areas worthy of protection (which is issue oriented and constantly changing) to celebrating the timeless beauty and complexity of nature's ecological web. The book's

Winter pine needles

emphasis now centers on what today's scientists are learning about old-growth forests. These investigations reach no further back than the 1970s and, in the words of Jerry Franklin, a leading forest researcher, came perilously close "to not happening in time." As editor and photo editor, our conviction is that the workings of the forest speak for themselves. If more people increased their awareness of these dynamics, society as a whole would treasure forests for what they *are,* not just for the wood they provide. In short, the more we humans know, the more we will care; and the more we care, the better will we *take* care.

That said, we also wanted these pages to be engaging as well as informative. We therefore sought out writers not only conversant with forests but respectfully in love with them, and we asked each to offer tastes of his/her own particular geopolitical division—northern California, Oregon, Washington, British Columbia, Southeast Alaska. We hoped for a letter-from-the-field quality and in the interest of presenting a variety of perspectives chose five distinct voices: a university professor, an environmental journalist, a poet/nature writer who formerly worked in the woods, a government administrator/biologist specializing in interpretation, and a naturalist/consultant/teacher. As editor, Ruth would shape, coordinate, and fine-tune the manuscripts—a task surprisingly more pleasant than initiating one's own drafts. As photo editor, Charles would consult colleagues and assemble illustrations.

Each chapter, of course, was to be specific to its region and would portray the green legacy of that state/province by leading the reader vicariously into the forest. We also wanted the book as a whole to highlight how society's decisions affect preservation, how forest research goes forward, and what its major conclusions are. For example, scientists now realize that fire, blowdown, flood, insect infestation, and other forms of disturbance play a valuable role within the forest, and this is beginning to affect forest management practices. They are also investigating forest realms hitherto largely ignored. One of these is the canopy stretching above our heads—the entire three-dimensional geometry of twigs and foliage and space, where the weight of lichens and other plants using the treetops as perches exceeds that of tree foliage, and where thousands of species of insects live out their entire cycles. Another realm that is being explored is beneath our feet. Here tree roots and fungi form crucial mychorrhizal linkages, some trees actually needing a series of fungal partners at various underground levels and stages of their lives. A separate part of this subsurface realm lies within the gravels under rivers and extends out from their banks beneath the forest floor—an area known as the "hyporheic zone." Scientists have discovered there a distinct community of organisms, most still unidentified, let alone understood.

Material such as this is pertinent to forests throughout the Northwest, but to avoid overlap within the book we divided topics between the five chapters. Thus, the preservation story is told through an account of California's Save-the-Redwoods League, with its orderly combination of private philanthropy matched by state funding, and also through the peaceful citizen protests in British Columbia, which resulted in the greatest mass arrests in Canadian history. The ecological function of fire is emphasized in the Oregon chapter, although it also is touched on in the California and Washington chapters and of course pertains to all forests. Salmon as a forest product are discussed at greater length in the British Columbia chapter than in the other chapters because of a continuous twenty-year study of the issue there. The Washington chapter opens with a visit to the unique redcedar grove on Long Island (south of Aberdeen); the British Columbia chapter tells of the post-glacial arrival of cedar along the coast as revealed by pollen recovered from bogs, a progression ranging from 6,000 years ago for northern Washington and the Fraser River Delta to only 1,000 years ago for the Queen Charlotte Islands (just south of the Alaska border). The Alaska chapter includes a section on how life recolonizes land as glaciers retreat; the Washington chapter depicts the return of life to Mount Saint Helens following the 1980 eruption.

Throughout, the book recognizes people as part of the ecosystem. Western culture has tended to view *Homo sapiens* as apart from the rest of nature, whereas in actuality we have always belonged to the whole, often dominantly so owing to our innate ability to think and our penchant for engendering change. Historians, anthropologists, and ecologists now realize that our human effects began long ago. Over countless centuries Native Americans shaped the environment in various ways. With fire they created openings in the forest to enhance the growth of root crops such as camas and improve grazing for deer and elk (and after the mid-1700s for horses as well). People in the Queen Charlotte Islands seem to have inadvertently brought shrews and deer mice to the islands as stowaways on board their cedar dugout canoes. At Hesquiat, along the outer coast of Vancouver Island, they deliberately planted cattail so that they would have it for weaving into mats. At Bella Coola, they transported salmon eggs and stocked streams that seemed suitable but lacked particular runs of fish.

We who came later brought an ever-increasing hunger for commodities, along with the technological ability to provide them. In little more than a single century and a half, we altered the forests far more radically than the early people did in all their many centuries and millennia. By decimating the beaver population east of the Cascades, fur trappers unwittingly altered ecosystems—the beavers had created forests on sites otherwise too dry for them. Settlers brought sheep, horses, and cattle, which nibbled and trampled the land, altering the vegetation wherever they roamed. The new human arrivals also brought dependence on agriculture, and an accompanying concept of forests primarily as a crop of timber. And they brought urbanization, which has turned vast acreages of forest into lawns and asphalt, a process that continues.

All such manipulation forms a major thread in the fabric of human involvement with the forest. Another thread—a new one—is extensive scientific research. A third thread—even newer—is the realization that the best things in life are no longer necessarily free. We must pay economically, socially, and politically to protect and restore the environment. This awareness has appeared in an astonishingly brief time. The Douglas-fir beneath which Washington's Native Americans and the territory's new government signed a treaty in December 1854 still stands (watch for it while southbound on I-5, a tall dead tree south of the Nisqually River, near the bluff). From that time until now has brought both the ravishment of forests and the beginning of a new pattern. For the good news of the 1990s is that scientific understanding of forests is increasingly being applied to their management. Furthermore, most of the remaining, publicly owned old growth in northern California, Oregon, and Washington has been set aside for protection by the 1994 creation of seven million acres of federal forest reserves and two and one-half million acres of riparian reserves. British Columbia is taking steps in the same direction, albeit somewhat haltingly. In Southeast Alaska, current prospects appear less promising. Land within the boundaries of Tongass National Forest is largely rock, ice, scrub, or peatland. Only a small percentage of the lowland forest there harbors trees big enough to be coveted both for cutting and for preserving, and that rarity in itself renders them simultaneously vulnerable and uniquely valuable as habitat, particularly through the long, snowy northern winter.

Environmental policies and regulations and the boundaries of protected reserves are of course the product of people's decisions, and other decisions can undo them. A pendulum's tendency is always to swing—although perhaps the metaphor is imperfect when applied to human affairs. Our swings sometimes seem of uneven length each side of center; a lengthening arc of folly may follow a shortening arc of wisdom.

For most of the last five millennia Northwest forests stood thick, tall, and extensive. Massive needleleaf conifers dominated these forests, many of them the largest and oldest of their kind. Tree after tree towered more than 200 feet overhead. The volume of wood was truly stupendous—and for that reason is mostly gone, a boon economically and in terms of lumber and fiber used by us all.

West of the mountains, year-round moderate temperatures prevail, wintertime included, and weather moving directly in from the ocean drenches the land with high levels of precipitation and brings pulses of fog or clouds that partially offset summer droughts. Great amounts of moisture are stored in the soil, and the massive trunks of redwood and Douglas-fir constitute huge reservoirs in themselves. Calculations are that a tree 250 feet tall holds more than a thousand gallons of water, enough to permit photosynthesis even on hot days in late summer when smaller trees have shut down. Because of this and other factors the conifers of the maritime Northwest experience little metabolic "downtime." However, the same species that produce giants in the coastal environment, may never get "big" in the drier inland climate where refilling their internal reservoirs would be impossible. Evaporative loss from the crown would be too great. A single, huge coast-side Douglas-fir carries an estimated sixty million needles, a voracious potential "thirst."

John Sawyer, author of this book's California chapter, likes to point out to his classes that trees are among the most massive organisms in today's world, but actually none of the Northwest's giants holds the record for either age or size. The chief contender for maximum-known mass seems to be a soil fungus in Michigan that sprawls across 1,500 acres, or perhaps another of the same genus—*Amarilla*—recently found near Mount Saint Helens in Washington. Probably an aspen grove in Utah's Wasatch Mountains should also be considered. It is a clone, an organism that has perpetuated itself by vegetative reproduction rather than by producing seeds. It currently has 47,000 living stems, cumulatively triple the mass of California's General Sherman tree, which is the largest single living giant sequoia.

As for great age, the world's oldest living trees are found outside the Northwest. They are 5,000-year-old bristlecone pines in the high desert mountains of California and Nevada. Certain shrubs are still older. A creosote bush in California's Mojave Desert near Barstow has been sending out shoots for more than 8,500 years, constantly perpetuating itself. It is half again the age of the oldest living bristlecone. It, too, is a clone. But even it does not hold the age record. A box huckleberry, also a cloning shrub, in the pine barrens of New Jersey is said to be 14,000 years old! In the Northwest individual spruce, cedar, and Douglas-fir live "only" 500 to 1,200 years, but that is venerable enough to satisfy our human hunger for something that endures. From northern California to Southeast Alaska, each forest region has its own story to tell.

—Ruth Kirk
Lacey, Washington

—Charles Mauzy
Issaquah, Washington

Northern California

By John O. Sawyer, Jr.

Dr. John Sawyer is a professor of ecology and plant taxonomy at Humboldt State University in northern California. His knowledge of plants began on a practical level: family interest in growing them for food.

John has hiked extensively in the redwood forests and the rugged Klamath region, boot leather joining scholarly research as his base of knowledge. His professional work ranges from serving as a field ecologist in Costa Rica and Thailand to helping produce keys to California's vascular plants to making the first discovery of subalpine fir in California. He has developed methods of enhancing rare plant populations during restoration projects and has worked with the U.S. Forest Service identifying potential Research Natural Areas and for the National Park Service surveying possible natural landmarks. His contributions to scholarly journals number by the score.

Redwoods and sword fern

Rhododendron brighten a redwood grove scarred by fire. Bark six inches to a foot thick usually shields redwood from fire, while the wood itself is too moist to burn readily, and the pitch is virtually non-flammable. Even so, over time fire often eats into trunks without setting entire trees ablaze.

Northern California

Redwood Ecology

Growing up in Chico, in the northern heart of California's Central Valley, I was familiar with redwood as a tree but not as a forest. Redwood was what grew in people's yards to shade gardens and porches. I am almost embarrassed to admit that not until returning to California from graduate school at Purdue University, in Indiana, did I actually see this great forest wonder despite having lived so close to it. I took my parents for a classic tourist drive among trees as much as 1,250 years old and 368 feet tall, the tallest in the world, but I remember no particular moment of awe or enlightenment. Indeed, my main memory is of driving through fog.

Six years after that trip, I joined the faculty at Humboldt State University (then College) as a botanist with a special interest in the growing field of ecology, the interface of life-forms with each other and with their physical and chemical environment. This job has put me solidly in redwood country, both geographically and professionally. Thirty years and multiple graduate students and field studies later, I now feel a certain irony in knowing that there are both far fewer of the original redwood trees and far greater public interest in them than when I first drove to see them, let alone when I was growing up ignorant of their extent and magnificence.

To my mind, Eureka serves as a dividing line between California's southern redwoods and northern redwoods. North of Eureka, the redwoods of Humboldt and Del Norte counties mingle along creeks with Sitka spruce, western hemlock, and western redcedar. So many plant species common much farther north are present in this part of the forest that old-growth admirers might feel as if they were in Washington or British Columbia with redwoods somehow added. But south of Eureka, hallmark "Northwest" species drop out, and along floodplain terraces redwoods perfect in form and with diameters measured in tens of feet soar skyward. For travelers, these stands are the absolute pinnacle of the redwood experience.

Floods rather than tree-toppling winds act as the major disturbance of forests here. An example came during the winter of 1955–56 when a flood uprooted more than 500 large trees and washed away fifty acres of an Eel River terrace, which was 10 percent of that particular flat. Viewing the magnitude of the disarray and thinking of the land that had gone, some observers blamed recent watershed logging, but they were in part mistaken. Floods would have come anyway. Cutting probably worsened the effects, but major flooding here far predates the arrival of saws. In fact, according to studies by University of California scientists, it occurs every sixty to seventy years. There have been fifteen major floods in the last thousand years.

Unique attributes allow redwood to flourish in such circumstances and largely account for the purity of its stands. The alluvium deposited by floods holds ample oxygen and is rich in nutrients—and redwood is adapted to the sudden delivery of these elements. Instead of being killed by the burial of roots and lower trunk, as is true for species such as Douglas-fir and grand fir, redwood simply sends new growth out into the flood deposits. Root systems become layered, a new set developing to match each new deposit, and a ring of fresh sprouts issues from the trunk where it rises above the new ground surface. Redwood is also adapted to flooding by having cones that open in winter, which is flood season. Seed viability is only about 5 percent, but given the tree's longevity, a few successful seedlings every 500 years or so are enough for stand replacement. Seedlings germinating on bare alluvium escape the competition typical of the forest floor and are less likely to be "damped-off" by fungi than is common among those getting their start in duff and fallen needles. Observers sometimes worry that they do not see many oncoming young redwoods, but they need not concern themselves.

In 1967 the Eel River again rampaged; my first trip to the southern redwood flats came three years later. Redwood seedlings covered the ground. So did oxalis, which readily endures flooding because its underground stolons quickly send shoots to the new surface. Sword fern too was beginning anew, the nutrient-rich alluvium a suitable medium for its spores. Above this green carpet stood huge redwoods. Ten years later I returned to the same location. Many seedlings had died, especially those in the shade of trees, but dense patches of young redwood stood in openings. Rooted in moist soil just above the water table and supplied with optimum light, these trees may grow as much as ten inches a year, almost double redwood's usual annual growth. Thirty years after the flood, a few trees are now clearly taller than the rest. They will be the giants of centuries to come.

OREGON
SISKIYOU MOUNTAINS
JEDEDIAH SMITH REDWOODS STATE PARK
DEL NORTE REDWOODS STATE PARK
REDWOOD NATIONAL PARK
PRAIRIE CREEK REDWOODS STATE PARK
BALD HILLS
Klamath River
MARBLE MOUNTAINS
SALMON MOUNTAINS
TRINITY ALPS
MOUNT SHASTA
CASCADE RANGE
5
Trinity River
Arcata
Eureka
NORTHERN CALIFORNIA
Redding
LASSEN VOLCANIC NATIONAL PARK
HUMBOLDT REDWOODS STATE PARK
Eel River
101
YOLLA BOLLY MOUNTAINS
NEVADA
SIERRA NEVADA
Laytonville
Chico
N
W
E
S
20
Lake Tahoe
0
75 miles

Such river and creek terraces host all of the tallest and most massive individual redwood trees, but they make up only a small part of the landscape. Most redwoods grow more modestly on slopes above their justly famous peers. Their immediate associates are Douglas-fir, grand fir, tanoak, madrone, and canyon live oak, and the redwood trees accent rather than dominate the hardwood canopy. Herbs and ferns are virtually absent. This slope forest of the southern redwood region feels more like central and southern California than the Northwest.

Oregon white oak ranges from northern California to southern British Columbia and tolerates relatively dry conditions. It grows in open places away from conifers that might shade and crowd it out. Previous widespread burning by Native Americans helped produce landscapes well suited to this oak.

In the redwoods north of Eureka, floods are less severe and lowlands less extensive than to the south. Fire, present in both forests, here becomes the most obvious form of ecological disturbance. Bark scars and blackening are readily noticeable in most redwood stands, and many individuals survive as "chimney trees" with only outer wood and bark still present, as "goose pens" with huge burned cavities at their bases, or as "cathedral trees" with huge stems ringing a burned stump, testimony to long-ago fires.

Redwood's adaptations allow it to survive even intense fires. Its bark is thick and does not ignite easily, and the sprouting that is common following floods also occurs after fire. A scorched trunk may produce a whole new set of canopy limbs: vibrant green twigs destined to become true branches grow from a stark black trunk. During field studies, students and I occasionally have found what we think to be seedlings only to realize on careful inspection that they are sprouts originating from buried branches or burls, or fallen logs. Any part of the tree is able to sprout. In fact, instances of vegetative reproduction are so common I sometimes wonder whether individual, genetically distinct redwoods ever die—maybe they just keep on sprouting.

Both northern and southern redwoods receive from forty to eighty inches of rain annually, depending on the particular location, although in the north the rainy season stretches out a bit longer than is typical for the southern forest. Summer temperatures in both forests rarely rise into the eighties; fog is common, and humidity stays high year-round. Vegetation seldom dries out, a situation not conducive to wildfire. Burns come only about every 300 years, such a long interval that fuel builds up on the forest floor between burns and increases the intensity of fires when they do come. The long fire-free intervals also mean that trees have centuries to grow before flames again sweep through the forest. That gives redwood stands time to develop many of the old-growth aspects familiar farther north, including immense amounts of wood lying on the ground and trees in all stages of life. As a graduate student, Bruce Bingham, now with the Redwood

Sciences Laboratory in Arcata, measured woody debris on the slopes east of Prairie Creek. In places he found the ground covered to a depth of about ten feet. The volume of this debris was calculated at 200 tons per acre and the mass at 3,700 cubic feet per acre, figures comparable to those for Douglas-fir forests farther north. Much of the forest floor was literally out of sight. Bingham found himself walking on a tangle of jack-strawed logs rather than on the ground, and once a professor accompanying him—not I—fell into the jumble leaving only his feet sticking out. The position not only lacked dignity but was difficult from the standpoint of rescue.

Visitors to Redwood National Park expect to see this lush forest, but in the Bald Hills section of the park, just a few miles east of the coastal redwoods, they find a very different scene. Here the woodlands of Oregon white oak and swards of California oatgrass and various perennial grasses are representative of a problem. When park ecologist Mary Kektner was charged with returning the Bald Hills area to presettlement "natural" conditions, she and the Humboldt State students working with her soon realized the complexity of the problem, to some extent because people have been part of the equation at least for centuries and maybe for millennia. When I first came to the university, I commonly saw young Douglas-fir trees growing under the oaks. According to studies at the time, they had become established soon after herds of domestic sheep were removed, but new work indicates that this was not the beginning of the Douglas-fir invasion. The trees were there much earlier but were apparently held in check by the Chilula Indians' use of fire, which by keeping the Douglas-fir out of the oaklands helped to assure continuation of highly valued acorn crops.

We are now beginning to recognize the extent of human manipulation of landscape and how long it has been going on—and with that understanding comes growing awareness that creating a park does not mean the land will take care of itself nor does it mean that we will be able to step into a timeless forest virtually unchanged since the Pleistocene. Ecologists used to think that if human effects were eliminated, the land would return to its natural state. Now we find that there is no simple definition of what is natural. Indeed, disturbance and patchiness constitute the norm, and people are part of the ecosystem.

The great age of Northwest trees impresses us: individual monarchs bridge time back to the days of our great, great, great-grandfathers. But this does not mean that they are remnants of a uniform, self-perpetuating "original" forest. Quite the contrary. Nature does not maintain a steady-state balance. Instead it goes forward from event to event, whether a shift in climate, a volcanic eruption, or a locally great flood, fire, or windstorm. Forests as a whole constantly tune their responses to such changes even though individual titanic trees may live through them. Some forests may have started during a wet period but now live during a warm period—or vice versa. Nothing has halted their growth, yet conditions may have changed enough that the same overall kind of forest would not develop if it started now.

This means that today's human activities are upsetting not so much a balanced system as a grand old forest legacy. If we eliminate our old growth, it is not likely to come back in quite the same way. Saving this legacy is going to require considerable effort and money, but the arguments for doing so are the same as those used to justify the effort and cost involved in saving ancient documents or paintings—they are an indispensable and irreplaceable part of our human heritage.

Saving the Redwoods

Many sources estimate the extent of original redwood forests at just under two million acres. About 10 percent of that area is now in federal and state parks, and about 60 percent of the land within their boundaries has been logged. This translates into only about 4 percent of the original forests still unmodified by logging, at first blush a seemingly small success in saving redwoods. However, put in terms of nearly 80,000 acres set aside, it is a notable achievement, especially considering that these high-value trees were once privately owned.

When I went off to Indiana for graduate school in 1959 I learned what the presettlement midwestern forests and prairies were like. Field trips to their remnants interested me but were also shocking to a kid who had spent summers roaming the California mountains. These scraps of natural Indiana vegetation were dots in a landscape of towns and farms. Could this also happen in California? When I made my first visit to the redwoods, I found gems like the Avenue of the Giants and the Rockefeller Forest set aside in perpetuity and officially named, but I also learned about controversial plans for building a freeway through Prairie Creek Redwoods State Park, a still-forested island in a sea of cutover lands. This proposed action bothered me more than the stories of logging outside the parks. I knew that the commercial redwood lands were privately owned and committed to intensive management but had assumed that the state parks would remain preserved pieces of presettlement California. Apparently not, or at least not sufficiently. The generosity of private donors and the public trust were about to be violated. A freeway would turn a substantial amount of supposedly preserved forest into asphalt, and it would flout the established principle of a good but winding highway threading a series of superb redwood preserves that were intended to be destinations in themselves, not mere incidentals to be rushed through on the way to somewhere else.

As it turned out, at Prairie Creek the freeway was built around the park rather than through it—and that has created its own problem. The road rests on unstable land, which is slumping into the park and affecting streams. At Humboldt Redwoods State Park, the freeway slashes through the heart of the park, tearing so wide an artificial opening that moisture conditions seem to have changed, apparently causing the tops of trees along the east side of the right-of-way to die. The faster road has of course brought more people to the redwoods, for freeways seem to virtually cause cars. But whether high-speed savoring of nature is feasible or justifiable is another matter.

Nature keeps rearranging itself. A major earthquake in geologically recent time contributed to changing the Klamath River's course, turning the flow of this stretch into a placid, lily-dotted pond. In historic time another earthquake spilled sand dunes out into the forest, burying the bases of trees.

Once ensconced at Humboldt State University, I became immersed in the details surrounding efforts to establish Redwood National Park. These involved a broad range of issues: Was industrial land being properly managed? Was clearcutting the best way to harvest? What is "park-quality" redwood forest? Should "nonpark-quality" forest be preserved or left uncut? Here was no simple Congressional action turning so-called virgin land into a national park.

The contrast between original forest in the state parks and the lands outside them was extreme. Along the Redwood Highway, motorists traveled through green cathedrals but upon emerging were greeted by burned stumps and bare ground. That land is now covered with dense stands of trees, but a look back at the photographic record of the 1960s makes it easy to understand people's outrage at the time. The era's mix of persistent clearcutting, Congressional hearings, parades, newspaper headlines, and strategy meetings became one more chapter in the century-long attempt to protect samples of redwood forest. It was a battle with two major conservation associations as its standard bearers: the Sierra Club and the Save-the-Redwoods League.

The Sierra Club was founded in the 1890s, and at first its energies were focused on reforming exploitative forestry practices in the Sierra Nevada. Gaining protection for groves of giant sequoia and establishing Yosemite as a preserve were early accomplishments. So was strongly influencing the conservation and preservation policies applied to federal land during Teddy Roosevelt's presidency (1901–1908). However, because nearly all redwood lands were privately owned and therefore not subject to federal policy, the newly created U.S. Forest Service and National Park Service could be only marginally involved in redwood issues. If the trees were to be protected, it would have to be through other means. By the end of the 1800s groves near San Francisco had been logged, and by the 1900s railroads and truck roads were bringing redwood from increasingly distant sources to city markets. California Redwood State Park, now called Big Basin Redwoods State Park, was financed in part with state money in 1901. Seven years later Congressman William Kent used his own money to purchase Muir Woods, near San Francisco, which he donated to the federal government. Along with these actions, came the beginnings of talk about a redwood national park.

A major, well-organized approach to preservation got underway in 1918 with the establishment of the Save-the-Redwoods League, a product of the same movement that had spawned the Sierra Club. Prominent professionals, members of the business community, academics, women's garden club members, and automobile association members joined. The league looked to philanthropy as a source of funds to purchase groves and rights-of-way along the Redwood Highway from northern California to Oregon. Additionally, they commissioned the renowned landscape architect Frederick Law Olmsted to survey northern counties for possible national park locations. His 1927 report ruled out redwood stands at the mouths of the Klamath River and Redwood Creek as too large and inaccessible. Olmsted recommended focusing instead on "park-quality" groves lining the river terraces served by the existing highway, U.S. 101. Soon the league had four acquisition projects:

- Bull Creek–Dyerville (now Humboldt Redwoods State Park)
- Prairie Creek (now Prairie Creek Redwoods State Park)
- Del Norte (now Del Norte Redwoods State Park)
- Mill Creek–Smith River (now Jedediah Smith Redwoods State Park)

In Prairie Creek, as elsewhere, logs and other forms of coarse woody debris create vital habitat. Fish and tadpoles rely on the riffles, glides, and quiet backwaters the debris produces. Aquatic invertebrates use the wood as attachment sites and as bridges when moving from life in water to life on land.

To assist the fund-raising, the California legislature passed a bond act authorizing the use of public money to match the private donations raised for the purchase of land for state parks. By 1928 more than half a million dollars of private money were in the league's coffers, and after that year each additional dollar raised was matched with public funds. As Susan Schrepfer points out in her book *The Fight to Save the Redwoods,* "The effort bridged the traditional charities of the nineteenth century and the public structures of the twentieth; it was both zealous in its privatism yet democratic in its spirit."

Because they had been successful with this method of preservation, the league did not support 1930s-to-1970s campaigns to create a national park. They considered the state park system a better mechanism for saving and managing redwoods. The Sierra Club disagreed and in the 1960s demanded a large, ecologically viable unit within the national park system. The club pointed out that chain saws and caterpillar tractors lead to fast and destructive logging, and that vast forests were rapidly becoming nothing but dimension lumber and memories. Also, the heavy flooding of 1955–56 had shown the frailty of an approach to redwood preservation that depended on superb stands strung along highways. Hundreds of trees in Rockefeller Forest, purchased at considerable cost, had simply vanished that winter, their "eternal" quality washed away in the swirling waters of just a day or two. With that experience painfully fresh in their minds, and the pending replacement of the two-lane Redwood Highway by a four-lane freeway, many in the Sierra Club argued that a federal park offered the only secure, long-term way to preserve redwoods.

Today's almost 80,000-acre Redwood National Park is the outcome of that conviction, the work of a multitude of committed people over a long time. In addition,

redwood state parks protect 150,000 acres along river flats spread through the range of the species. Many of the nation's major kinds of vegetation have fared less well—including the forests and prairies of Indiana.

Beyond the Redwoods

Beyond the redwoods, travelers in northern California find forests of Douglas-fir, tanoak, and madrone alternating with Oregon white oak woodlands and grasslands. Books refer to this vegetation in various ways: Douglas-fir–hardwood forest; mixed evergreen forest; Douglas-fir forest on mixed hardwood sites. Whatever called, it is a forest that extends from the coast hills of central California into the lowland slopes and canyons of the California-Oregon border. Almost all of this forest has been extensively altered by logging, which intensified after World War II when returning GIs created a great need for houses. At the same time the state tax on standing timber gave ranchers an incentive to change trees into cash and maybe get more grazing land in the deal. The independent loggers ("gyppos") who cut the Douglas-fir tended to leave the hardwoods, and that has resulted in increased tanoak and madrone in the forest today.

In this mostly logged region, the Keith Angelo Coast Range Preserve west of Laytonville remains as an extensive uncut area. The Nature Conservancy's first preserve in the state, now part of the University of California preserve system, it holds old-growth forests of widely spaced Douglas-fir 200 and more years old in combination with tanoak and madrone half their height. These stands are mixed with others of younger age and with chaparral and grasslands. The old growth is a place to hear spotted owls calling in the evening and to find mammals from deer and bobcats to red tree voles. But it is not a forest to mistake for those of Oregon, Washington, and British Columbia. No western hemlock or redcedar, Sitka spruce or grand fir join the Douglas-fir here in neck-craning array. This forest, just south of the Redwood Highway, is like the slope forests of Humboldt Redwoods State Park but without the redwood.

East of the Redwood Highway in the Klamath Mountains of the California-Oregon borderlands, one can see forests less modified by human activity than is true in other parts of California. Intensive U.S. Forest Service management, hence logging, did not begin there until the 1960s. Ancient trees are few because fire comes too often to much of the area. Its intervals are spaced by decades instead of centuries, a frequency that points to relatively cool ground fires rather than high-temperature, stand-clearing fires. This has led to patchiness, with forest in all stages of development, rather than extensive stands of old growth. The patches, however, maintain the greatest diversity of plant species anywhere in the Northwest and are remarkable even on a world scale.

The region fascinates botanists and ecologists. Over the years James Smith (also of Humboldt State University) and I have made hundreds of field trips to find out just how great the plant diversity is. We now count over 3,500 taxa in 150 families and 760 genera in the region north of State Route 20 and west of the I-5 freeway. Among these plants are an astonishingly high number of endemics: 281 species, subspecies, and varieties that grow here but nowhere else. Why? David Rains Wallace in his book *The Klamath Knot* attempts an explanation by pointing out that three mountain ranges, each with its own characteristic vegetation, converge in this border region: the Sierra Nevada, Cascade, and Coast Ranges of California and Oregon. Robert Whittaker, a leading plant ecologist, recognizes an even greater convergence of plants, with species from the Pacific Northwest growing along with those of the Southwest. From each direction, many have ranged this far and stopped. The Klamath region is a meeting ground. Conifers provide my favorite example of this phenomenon. Yellow-cedar, Pacific silver fir, subalpine fir, Engelmann spruce, and noble fir grow no farther south than the Klamath region; and foxtail pine, gray pine, and knobcone pine reach no farther north. That makes eight species. To them add the nineteen tree species common throughout these mountains plus two found only here, Brewer spruce and Port-Orford-cedar. This swells the total for the region to twenty-nine kinds of conifers, a rich variety.

Sundew are carnivorous plants that grow in bogs and wet meadows. Their leaves bristle with stalked glands dotted with droplets of "glue." Small insects which touch the glands become ensnared when the stalks bend down over them and are then digested by enzymes.

The Klamath region is almost completely mountainous, a great upturned land where rivers create sweeping arcs, cutting across bedrock on their way to the sea. Human development is almost an anomaly. Few towns even have stoplights. Instead of cities and freeways, officially designated wildernesses are the mode here. Their names are pleasing: Kalmiopsis, Rogue River, Red Buttes, Siskiyous, Marble Mountains, Russian, Trinity Alps, Canchian, Yolla Bolly.

At just over 9,000 feet, Mount Eddy, on the eastern border with the Cascades, is the highest peak in a veritable maze of ranges that merge one into another. However, in spite of only modest heights, these mountains are rugged. Steep slopes cut by canyons characterize their lower elevations. At 4,000 feet the pitch becomes more gentle, but by 6,000 feet it steepens again into glaciated terrain. It is diverse topography with complicated geology and as a result complicated botany. The rocks are old; they have been available to terrestrial plants throughout the Cenozoic Era, the last sixty-three million years. During this time mountains in this region have risen and worn down at least twice, but neither volcanic flows nor glaciers have ever completely covered them. Plants have always had places to grow, and their relation with the underlying rock has been

A pine grows from a rock crevice above Lower Caribou Lake in the Trinity Alps. The Trinity, Salmon, Marble, and Siskiyou Mountains make up what is called the Klamath Mountains, or Klamath region. Rock here is older than in the Coast Ranges and the Cascades. The region dates to thirty million years ago when a low, rolling plain stretched across most of today's American West.

extraordinary. Much of the region is composed of metasedimentary rock mixed with outcrops of granitic and ultramafic rock, and in some local areas, marble. Botanists call the ultramafics and the soils that develop from them "serpentines." They notice the shiny green color of the rock, which weathers to a brick red soil. Even more they notice the special plants associated with them.

Ultramafic rocks are characterized by high levels of various metals and are odd in that they can vary substantially in type within a short distance, a variation that is accompanied by abrupt changes in vegetation. The most extreme of these changes comes at contacts between serpentine and sedimentary soils. On one side of the junction, the soil is gray (it may have developed from sediments) and the forest is composed of dense and massive firs. A few feet away red soils support sparse stands of short pines—sugar pine, western white pine, lodgepole, knobcone, or Jeffrey—which species grows on what particular site depends on the specifics of that location. Most vegetation on serpentine soil is stunted in appearance and looks infertile. Even in this region where precipitation totals over 100 inches annually, about triple that of Seattle, plants look as if they need more water. Here, "big" equates with reaching shrub size. Leaves are evergreen, leathery, and small. Members of the heather family tend to dominate. What trees are present—mostly pines—grow scattered across extensive grassy areas with perennial herbs intermixed. In some areas, certain soil conditions result in barrens virtually without woody plants. Oddly enough, vegetation on Klamath serpentine may even resemble that of northern bogs. Primarily this is because both serpentine and bog habitats supply low levels of nutrients such as nitrogen, phosphorus, potassium, and sulfur, and the overall sparseness of the vegetation lessens nutrient cycling. To this, serpentines add features all their own, which challenge plant growth still more. Soils are almost always shallow and may have high concentrations of chromium, cobalt, copper, iron, magnesium, nickel, and zinc, which tend to be toxic.

Klamath plants long ago adapted to such conditions. Some grow only on serpentine soils, others on nonserpentine soils as well. University of Washington botanist Arthur Kruckeberg finds that the widespread species probably adapted to the rigors of serpentine life by developing physiologically distinct races. Varied soil types would be expected to support more plant species than if there were only one kind of soil, and varied soils have been present here for millennia. Plants have had time to adapt.

Two other factors also add to the diversity. The Klamath region has a long history of local disturbance, and regardless of our human tendency to see disturbance as loss it may actually give rise to diversity. Fires, floods, insects, and pathogens are part of the flux here, as they are throughout Northwest forests. In addition, this topography is so rugged that in places elevation varies by 5,000 feet within a single square mile. No extensive same-composition, same-age forests have developed. Habitats vary too much. A hiker may encounter changing rock types several times in an hour or two and see stands of trees in all stages of development, chaparral, meadows, and rock outcrops. Ecologists talk about "high habitat heterogeneity" as a way of summarizing this. One particular square mile I know of has seventeen kinds of conifers, almost two-thirds of the number for the entire region. This level of diversity is astonishing within a square-mile patch of a region as small as the Klamath, especially in the context of a land renowned for its vast expanses of Douglas-fir and fir forests. The Klamath is like a sampler: a little of this, a little of that, all mixed together.

Almost a century ago the U.S. Biological Survey's revered biologist C. Hart Merriam pointed out the influence of elevation and latitude on plant life. He noted that high elevations in Arizona, clothed with spruce and fir, resemble the forests of high latitudes of Canada. Ecologists no longer use his terminology, but the pattern he first called attention to remains: vegetation varies with both elevation and latitude. This phenomenon is now expressed in the concept of zones, which are defined locally for each region or mountain complex. Zones include consideration of each area's specific characteristics as well as its elevation and latitude—and certainly the Klamath region has a plethora of specific characteristics. It demonstrates the overall principle of zones, then adds its own peculiarities. For example, the Siskiyous and the western face of the Klamath Mountains in California host Douglas-fir, tanoak, and madrone at low elevations, and in Oregon those three species are joined by two more: grand fir and western hemlock. The low-elevation Trinity Alps and the eastern face of the Klamath Mountains present a different picture. There, ponderosa pine joins Douglas-fir as the main species.

The same sorts of variation also apply to the montane and subalpine zones, but far from exclusively so. Serpentine vegetation follows its own course. Douglas-fir—so extensive elsewhere—is almost missing here at lower elevations. Instead, knobcone pine, Jeffrey pine, western white pine, and incense-cedar are common, and at high elevations they are joined by foxtail pine, a tree that grows only in the southern Sierra Nevada and the Klamath-Siskiyou country.

As a botanist roaming these mountains, I have pondered these peculiarities and also puzzled over the great number of endemic species such as the azalea-like kalmiopsis and Brewer spruce. I am not alone in this. Fifty years ago Ledyard Stebbins, geneticist and

Leathery leaves and self-shredding, peeling bark set madrone apart from other Northwest trees. It needs mild winters but can thrive with an annual precipitation as drenching as 150 inches or as stingy as 15 inches. Madrone ranges from California to British Columbia.

Pitcher plants reverse the generality that plants nourish animals; growing in poor soil they get nitrogen from insects. A sticky fluid held in the plant's tubular leaves drowns the insects, which then are digested by bacteria, making their nutrients available to the plants.

evolutionist at the University of California, partially explained the situation by pointing out that there are two categories of endemics, roughly defined as old ones and new ones. The old ones, "paleoendemics," are derived from ancient stock. Their closest relatives are now either fossils or at least are not growing in nearby regions. For example, the closest relatives of kalmiopsis are now in Europe. Brewer spruce, another paleoendemic, is not related to either Sitka or Engelmann spruce, both nearby, but to spruces in Japan and Mexico. Sadler oak is related to an oak in Japan. The local pitcher plant is related to a species in the eastern United States.

Botanists expect to find large numbers of long-established endemics in an ancient landscape, but the Klamath region is unusual in also having many "neoendemic" species, newly evolved but with close relatives fairly nearby. A joking—but effective—way to find concentrations of these plants is to seek places where botantists disagree over what to classify as a species, what as a subspecies. That in itself suggests currently active genetic change, and it is definitely a situation that applies to this region. Many Klamath genera readily come to mind—*Sidalcea, Sedum, Lewisia,* and others. Most are candidates for consideration as neoendemics. All have numerous species that are "hard to tell apart."

I sometimes speculate that a third characteristic, easily overlooked, may also add to the mix here: climatic changes. Emphasis on the extreme age of the region and its varied plant cover so absorbs botanists that it is easy to forget about the swings of climate. Yet droughts have alternated with drenching rains, warm centuries with cold centuries, and this quite possibly has had a peculiar effect here. In other regions climatic changes may cause species to go extinct; their fossilized leaves, twigs, and seeds remain in ancient wetland depressions and swamps, but the plants themselves may no longer be present. Here, however, the great mosaic of habitats seems to have let established plants find suitable niches and continue growing despite major environmental changes. To these species are added new arrivals suited to the new conditions.

Subalpine fir, one of my favorite trees, may illustrate the arrival of a species. Dale Thornburgh, professor of forestry at Humboldt State, and I were the first to discover this fir in California. Classically Christmas-tree-like in silhouette and familiar in mountains from Oregon to central British Columbia (and at sea level in cold-air drainages north of there), the tree now is known at five locations in California. We first found it growing in the Salmon Mountains at Little Duck Lake, a glacial tarn created during the last ice age. Ed Cope, one of my graduate students, sought to discover where it came from. He compared the chemical makeup of subalpine fir needles at Little Duck Lake with some from near Crater Lake, Oregon, a close-by population, and also from the Colorado Rockies, a distant population. He found that tissue from Little Duck firs resembles that from Crater

Lake trees. David deJagen, a University of California graduate student, speculates that the subalpine fir may have immigrated here recently. Based on the age and structure of fir stands in both regions, he gives the tree's arrival at Little Duck Lake as within the last 2,000 years—indeed a newcomer. Other botanists speculate that the fir may be relicts, holdovers somehow surviving from an earlier time. One species, two possibilities.

Might the Klamath region "collect" new species and rarely lose them? This may be a key to its acclaimed plant diversity: not just varied soils from varied rocks, which lead to endemics, but also plants continually arriving from outside the region and not leaving.

The Klamath legacy is unique and relatively intact. The region has no cities or major agriculture. Placer mining and dredging during the gold-rush years left scars, but for the most part these have softened. Recent logging is common but limited to rather small swaths. Cows and sheep graze forests but no longer in the decidedly destructive numbers that were typical earlier. And offsetting these impacts are large tracts of relatively unchanged lands now officially designated as wilderness. The combination provides possibilities for maintaining the Klamath's diversity. Large reserves are still feasible here, and they can be interconnected. New forestry techniques and controlled grazing can be practiced on managed lands. Unlike so many other places, we do not have to settle for remnants of legacy. We can perpetuate the full original mosaic, or at least not threaten it with our human activities. We can allow the land to respond on its own to the vagaries of time and change.

Oregon

By Kathie Durbin

Kathie Durbin, granddaughter of a logger, was born and raised in Eugene, Oregon, where she grew up surrounded by the Willamette National Forest. Her father took her trout fishing and her mother, who taught wildflower identification and haiku to the kids of a timber town, imbued her with a love of words and nature.

Kathie has worked in Oregon as a journalist since 1974, for five years covering all aspects of the Northwest old-growth forest conflict for *The Portland Oregonian*. Since 1994 she has been writing about environmental issues for *High Country News, National Wildlife, Audubon,* and other periodicals. Her book *Tree Huggers: Victory, Defeat, and Renewal in the Northwest Ancient Forest Campaign* explores the history and impact of the citizen forest-preservation movement.

Right: *Kalmiopsis Wilderness*

The rare Brewer spruce is also called weeping spruce because of its dangling, stringy branchlets. Once fairly widely distributed, the tree now grows as an endemic species native only to the California-Oregon borderlands. It prefers north-facing slopes where snow, hence moisture, lasts into summer and fire danger is relatively minor.

Oregon

The Siskiyous: Ecological Crossroads

The fire-gutted incense-cedar known as Foster's Temple rises at the edge of a mountain meadow in the Red Buttes Wilderness, in the Siskiyou Mountains a mile or two north of the Oregon-California border. The south-facing trunk of this thousand-year-old behemoth, a gnarled stem weathered to shades of rust and gray, gives no hint of its ravaged north side. Fire charred the tree many centuries ago. A person can now stand upright in the gnome's house of its hollow base, looking out through a slit and viewing the pink and yellow wildflowers that embroider the surrounding meadow. Small, symmetrical incense-cedar seedlings stand like acolytes at the base of the great tree. Its sturdy elephant toes look like they must dig into the earth, anchoring the massive stem, although actually incense-cedars are shallow-rooted.

I have come to Red Buttes in a spirit of obeisance and because monarchs like this monster cedar are part of my beat as an environmental writer. Since the late 1980s, old growth has become a public issue as well as a forest legacy, and in its 20,000 acres this seldom-visited wilderness reserve harbors venerable specimens of old-growth sugar pine, Douglas-fir, grand and noble fir, and Port-Orford-cedar as well as incense-cedar. Red Buttes, named for twin cones of red peridotite rock, is a remote botanical gem, one of several in this mountainous crossroads of climates, biological communities, soils, and topographies. At Soda Mountain, east of Ashland, western juniper from the Great Basin desert mingle with typical coast conifers. To the west, near Cave Junction, the serpentine soils of Eight-Dollar Mountain support scores of endemic plants, including two dozen species listed as endangered, rare, threatened, or sensitive; no other site in Oregon has so many species of concern—and this is a state blanketed with forests of stunning diversity.

Southwest of Grants Pass, the Kalmiopsis Wilderness Area hosts the rare azalea-like shrub for which the wilderness is named, *Kalmiopsis leachiana.* At the eastern edge of this wilderness, a trail leads to a stand of otherworldly Jeffrey pine. Very old but not very big, these gnarled trees are shaped by wind and by the mineral-laden serpentine soils of the Siskiyous. They seem to crawl along the ground and grow back upon themselves, interspersed with pinemat manzanita, prostrate juniper, and ceanothus. Ten minutes by trail from the pines, rare Brewer spruce trees droop their weepy branches. I once met a couple from England there. Familiar with the tree in English gardens, they had come all this way to view it in its natural home. Across from the spruce, my guide Barbara Ullian—who never tires of photographing and botanizing this area—pointed out a Bolander's lily, with its hanging, trumpet-shaped bloom, and a rare phantom orchid, white and ghostly, rising from the duff of the forest floor. Both plants are Siskiyou endemics, common here but found nowhere else.

Throughout the Siskiyous, and all of the Klamath region—as elsewhere—wildfire has been a major shaper of the forest. West of Grants Pass, scars on old trees testify to large burns of more than 300 acres coming at intervals as frequent as every fifty years. One ponderosa pine bears eleven scars from fires in the sixteen and a half decades between 1814 and 1980. Yet the frequency, extent, and intensity of fires have varied; some 190-year-old trees show no fire scars at all.

Most wildfires are ignited by lightning. That was the case with the Silver Complex fire, which in 1987 swept across 96,000 acres of the Kalmiopsis Wilderness and the North Kalmiopsis Roadless Area, scorching ridgelines, dipping into river canyons, and leaving untouched green stringers of conifers clearly visible from the air. The U.S. Forest Service has tried to emulate this pattern in its design of some recent timber sales, showing a newfound respect for fire's role in the ever-changing, ever-renewing life of the forest. But for centuries—probably even millennia—indigenous people here understood the role of fire and used it to encourage many of the plants they relied on for food and medicines, and those that nourished deer and other game animals. No one knows how frequently they set these fires or how big they were, but Dennis Martinez of the Takelma Intertribal Project based in the town of Talent, near Medford, argues that "people have been key players in ecosystem dynamics for a long time in the Pacific Northwest—no less so than a key pollinator or carnivore or any other indicator species." On this basis, Martinez is working to get 17,000 acres of the Rogue River National Forest set aside for the practice

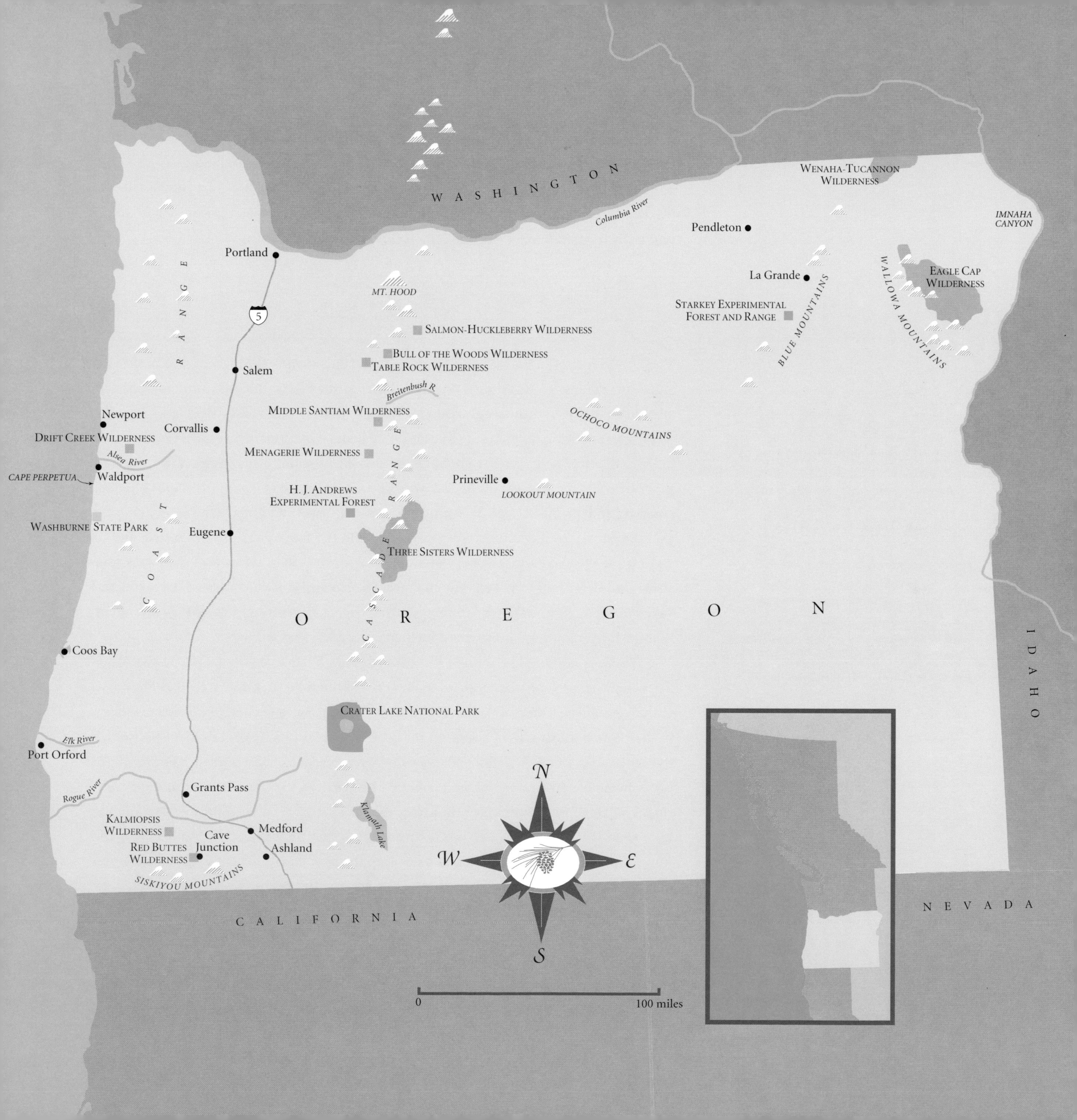
WASHINGTON
Columbia River
WENAHA-TUCANNON WILDERNESS
IMNAHA CANYON
Pendleton
Portland
La Grande
EAGLE CAP WILDERNESS
WALLOWA MOUNTAINS
MT. HOOD
STARKEY EXPERIMENTAL FOREST AND RANGE
BLUE MOUNTAINS
5
SALMON-HUCKLEBERRY WILDERNESS
BULL OF THE WOODS WILDERNESS
TABLE ROCK WILDERNESS
Salem
RANGE
Breitenbush R
MIDDLE SANTIAM WILDERNESS
OCHOCO MOUNTAINS
Newport
Corvallis
DRIFT CREEK WILDERNESS
Alsea River
MENAGERIE WILDERNESS
CAPE PERPETUA
Waldport
Prineville
LOOKOUT MOUNTAIN
H. J. ANDREWS EXPERIMENTAL FOREST
WASHBURNE STATE PARK
Eugene
THREE SISTERS WILDERNESS
COAST
CASCADE RANGE
OREGON
IDAHO
Coos Bay
CRATER LAKE NATIONAL PARK
Elk River
Port Orford
Grants Pass
Rogue River
Klamath Lake
KALMIOPSIS WILDERNESS
Medford
Cave Junction
RED BUTTES WILDERNESS
Ashland
SISKIYOU MOUNTAINS
N
W
E
S
CALIFORNIA
NEVADA
0
100 miles

of indigenous forestry, including deliberate burning to restore native oak savannas and enhance their production of traditional food and medicine plants and crops.

"Ecological stability has a human dimension because we are part of the natural world," Martinez comments. "Western culture is caught between environmental abuse and static preservation. Neither posture is sustainable or sustaining."

To the Sea

Silvery drift logs at Bandon are typical of those along the entire Northwest coast. Sand accumulates behind such logs, in some places giving birth to dunes, in others stablizing low-lying shores and providing habitat for plants and a range of animals from insects and birds to chipmunks, mice, skunks, and raccoons.

The Elk River rises in the Siskiyou Mountains and cuts steeply through rugged peaks and ridges on the way to its estuary, near the coast town of Port Orford. Conifers overhang and shade the river, drop needles into its swirling waters, and anchor the banks, keeping soil from slipping in and burying the spawning beds of salmon. When the trees die, some fall into the river and form the pools, riffles, and shallows that fish require. This river and other free-flowing streams of the Siskiyous and Coast Range still support scores of coho, chinook, and steelhead runs. To stand beside such a stream within a forest is to recognize that the fate of Pacific salmon is synonymous with the fate of old-growth forests and to feel intuitively the connectedness of water and land. Beaver dams and drift logs give shape and definition to the stream channel. The stream in turn carves and recarves the valley through which it flows, eroding, shifting course, depositing rich alluvial sediments that nurture grasses and herbs, watering bankside alders and willows.

"There is nothing like the Elk River North Fork fishery anywhere else in the lower forty-eight states," proclaims Gordon Reeves, a Forest Service research fish biologist. Intact coastal watersheds are virtually gone in Oregon, but a few retain a pristine quality in their lower reaches. Drift Creek on the central coast is one, the Salmonberry River on the north coast another, and the Elk River here on the south coast a third. The trees have been shorn from the ridge tops and tributary headwaters, the upper slopes, and the mid-slopes of the upper Elk drainage, in places even near stream banks. But downstream, the watershed is largely intact and productive enough that biologists come to decipher the forest's role in shading, feeding, and harboring the river's magnificent chinook—salmon famed among sportsmen for their size, aggressiveness, and bright flash in fast water.

Reeves and others with the Forest Service's Pacific Northwest Research Station, the Oregon Department of Fish and Wildlife, and Oregon State University have studied not only the river's remarkable fishery but what threatens it. Landslides are a natural phenomenon in the Coast Range, where highly erodible, supersaturated soils periodically separate from underlying bedrock, triggering mass land movements that can wipe out

spawning beds. Logging—especially its road building—often exacerbates this tendency, as Siuslaw National Forest officials discovered in 1975. That year, in just thirty-six hours, one rainstorm totaling 7.76 inches soaked the ground and set off 245 slope failures, all but 27 of them traceable to logging and road building. Habitat for countless salmon and steelhead was ruined.

As a result, the agency developed new logging standards to protect the headwalls of coastal streams from the consequences of logging and for the disposal of dirt from road building in ways that would not increase the risk of slides. Other federal and state agencies have now followed suit. Their new standards focus on limiting, or prohibiting, logging within 100 to 300 feet of stream banks. But as a number of studies document, it is

Trillium and bleeding heart garland the base of a cedar tree. Trillium seeds have oily appendages that attract foraging ants, a means of assuring dispersal for the plant. The ants carry off the seeds but eat only the appendages; what remains readily germinates.

not just conditions within adjacent riparian zones that affect streams. Upland areas are critical as well. Everything runs downhill and, eventually, downstream.

Studies have also found that a great amount of woody debris occurs naturally in healthy streams. Nineteenth-century fur trappers and settlers noted that streams in western Oregon and Washington were usually clogged with wood, impeding navigation. For instance, in 1870 logjams forced the Lower Willamette into five separate channels between Eugene and Corvallis. According to reports, in one ten-year period men pulled more than 5,500 logs measuring five to nine feet in diameter from a fifty-mile stretch of the river. Bearing out this abundance of wood in streams, present-day researchers from the U.S. Forest Service, the Weyerhaeuser Corporation, and Oregon State University—Jim Sedell, Fred Swanson, Peter Bisson, and Stan Gregory—found that small streams may contain as much as 700 tons of woody debris per acre. Loggers formerly added to it, dumping in excessive amounts of debris, which depleted oxygen in the water, smothered fish habitat, and formed temporary debris dams. Recognizing the damage this caused, federal and state forestry agencies began to require removal of all debris from streams and their banks after logging. Ironically, however, this clearing left streambeds vulnerable to scouring during periods of high runoff. With greater understanding of the role wood plays, agencies now require that coarse woody debris be left in the water, and even that some previously removed be put back. Also, live trees and dead snags along banks are not to be cut—and not just because they provide shade. Over time they will replenish the supply of debris in the water.

The loss of shade and forest structure, together with the increased sedimentation of streambeds when logging occurs near rivers, have all contributed to the decline of Pacific salmon. The American Fisheries Society warned in a 1992 report, *Pacific Salmon at the Crossroads,* that at least 214 salmon runs in Washington, Oregon, and California were at risk of extinction because of habitat degradation and other factors. In the late 1980s, concerned about the Forest Service's plan for further logging in the Elk River watershed, a coalition of environmental groups, Native American tribes, fishermen's organizations, and seafood companies proposed an Elk River Salmon Conservation Area where only minimal timber harvest would be permitted. The protective legislation never happened, but as concern for the salmon's plight heightened in the mid-1990s, President Clinton endorsed a westside forest plan for the Pacific Northwest built on the concept of protecting 164 key watersheds in western Oregon, western Washington, and northwestern California through sensitive forest management.

The effects of the plan reach to the Pacific itself, because estuaries, with their mingling of fresh- and saltwater, are linked to the forest. In a study released in 1988, H. J.

Andrews Experimental Forest researchers Chris Maser, Robert Tarrant, Jim Trappe, and Jerry Franklin explored the role of coastal drift logs in stabilizing dune fronts and creating shaded microclimates where plants can colonize sand. They found that stranded driftwood captures sediments and gravel, as well as directly shielding the shore from wave action. They also found that Sitka spruce and huckleberry take root and grow directly on logs strewn about estuaries, and that birds such as crows, gulls, eagles, and kingfishers use them as resting and hunting perches.

Furthermore, forest effects do not stop at the edge of the saltwater. While afloat along rocky shores, storm-driven logs alter the marine community by opening patches in the seaweeds, barnacles, and mussels that cover intertidal rocks, preparing sites for recolonization and starting the chain of organisms anew. In the open sea, floating logs provide movable habitat for gooseneck barnacles and attract schools of tuna, perhaps because of the shadow they cast. Sunk to the ocean floor, they serve as a base for various other life-forms, and these attract schools of small fish, which attract larger fish, and so on. In time, the logs and other woody debris rot and themselves enter the ocean's food chain—which supports the salmon, which swim back upriver and return to the forest. Linkages.

Oregon Coast Range: Fragments and Mosaics

I am driving the winding Alsea Highway through the Oregon Coast Range, threading green valleys where Oregon white oak cloaked with lichens screens the Alsea River, and apple trees loaded with blossoms glow in faint sunshine. The road enters a tunnel shaded by the graceful boughs of giant Douglas-fir, then climbs into the high country, where the view opens to the shoulders of clearcut mountainsides now coming back to fireweed, huckleberry, and salal. Most of these gentle hills are owned by timber companies; most have been logged twice in the past century. Because of that, and also because of cataclysmic mid-1800s fires that destroyed immense swaths of forest, true old growth is scarce in the Oregon Coast Range. Yet mild climate and heavy rainfall provide such ideal growing conditions here that stands of conifers 120 years old give the impression of being a century older. Many have already acquired old-growth structure—large trees, a multilayered canopy, an open interior forest.

A sudden rain squall hits. I pass through it and soon am on a logging road heading for the Harris Ranch trailhead, my entry to the Drift Creek Wilderness, heart of the largest protected block of unlogged forest left in the Coast Range. Only about 6,000 acres of the watershed are officially dedicated to the present and future as wilderness, and about twice that acreage remains in a near-natural state.

The largest trees—western redcedar and bigleaf maple—grow far below in the valley bottom. Here, the steep slopes are thickly forested with tall Douglas-fir, western hemlock, and redcedar, and a classic Coast Range understory of salmonberry, huckleberry, salal, and Oregon-grape. Today the forest floor is punctuated with the virginal white triple petals of trillium, and as I slip and slide down the muddy trail I also notice yellow wood violet and the pink-lavender teardrops of bleeding heart. Nurse logs and the thick duff of the forest floor nurture a terrarium of tiny hemlock, huckleberry, and unfurling sword fern. On the forest floor, prostrate conifers are doing their work as seedbed for tomorrow's forest. As their roots wrench free, the great trees' fall has already mixed rich organic topsoil with underlying mineral soil, a process called "root plowing." This process creates a rich medium for the next generation of trees and for the plants that will spring up while the trees are growing tall. I pause at an opening where the opposite ridge is visible through rising mist. It is a rare vista so near the ocean: an intact forested watershed, unbroken from rim to rim. I breathe in the damp clean air, hear the call of unseen birds overhead, feel my spirit lift.

The Oregon Coast Range encompasses about five and a half million acres of highly productive forests that stretch from the Coquille River north to the mouth of the Columbia, west to the Pacific headlands, and east to the edge of the fertile Willamette Valley. Gentle terrain, mild climate, and enormous trees attracted nineteenth-century timber barons, who acquired the most accessible land and over the next century systematically cut the forest, resulting in today's mix of fresh clearcuts and plantations.

Topographically diverse, these mountains climb from sea level to just over 4,000 feet. Close to the Pacific shore, where rainfall averages 70 to 100 inches annually, loamy soil supports a Sitka-spruce rain forest. In the late 1980s this forest drew the attention of the Washington, D.C.-based environmental group Conservation International, which undertook computer mapping to determine the extent of temperate rain forest worldwide. The results indicated how scarce this forest actually is, occurring only in latitudes from thirty-two to sixty degrees north in a narrow band between ocean and coastal mountains. Rainfall there exceeds 80 inches annually and catastrophic wildfire is infrequent, in marked contrast to conditions farther inland.

For the contiguous forty-eight states, the mapping revealed that the largest intact representation of rain forest is a 33,000-acre remnant on the central Oregon Coast between Cape Perpetua and Washburne State Park. Most of this lies within two watersheds, administered as wildernesses by the Siuslaw National Forest, with a finger of private land sandwiched between them along Tenmile Creek. The crucial value of the

A ground fire may suddenly flare into the crown of a tree, creating pillars of flame that set nearby trees ablaze. Ecologists now recognize the role of wildfire in perpetuating forests on both sides of the Cascades.

two separated wildernesses prompted the National Audubon Society to buy a particularly critical tract of the private land as a link for wildlife moving from one federally protected watershed to the other.

A few miles east of the narrow band of Sitka-spruce forest, the vigorous, fast-growing Coast Range mix of western hemlock, western redcedar, and Douglas-fir begins. Mist, rain, wind, and temperate climate shape this forest, home to deer and elk, black bear and cougar, spotted owls and marbled murrelets. Red alder, bigleaf maple, Oregon ash, and black cottonwood grow along streams, and at the forest's southern end savannas dotted with Oregon white oak provide a transition to the arid Umpqua River Valley, famed for spring wildflowers.

This forest, too—like virtually all Northwest forests—has been influenced by fire. Indeed, major stand-replacing fires have apparently occurred in much of this area every 400 to 500 years. For most such fires this is too long ago for the causes to be known, but within historic time, the causes of these conflagrations are evident. In the late 1800s a fire, apparently touched off by Native Americans to improve browse conditions for deer, instead raged into an inferno fanned by dry east winds. This was the Yaquina Fire, which

burned a total of nearly a million acres between Newport and Corvallis, smoldering and flaring for month after month before it finally went out. In 1886 the Coos Bay fires burned 300,000 acres as a result of careless logging practices. A half century later the Bandon Fire, also caused by logging, burned 144,000 acres. In that area, tree-trunk scars indicate fire intervals of 90 to 150 years.

In the case of the well-known Tillamook Burn, three devastating fires that came between 1933 and 1945 swept across part of the Coast Range, burning a total of 355,000 acres. Thanks to a reforestation campaign that enlisted penitentiary inmates and Portland schoolchildren, roughly half of this land was replanted with Douglas-fir seedlings, one of the first massive hand plantings ever undertaken. The other half of the land was seeded from an airplane with Douglas-fir. Today the trees are approaching harvest age, and the view from high ridges opens on green stands that stretch as far as the eye can see: a success. Had the burn been left to reseed itself, it would have come back in much this same way, mainly in Douglas-fir, a species that thrives in open conditions but is intolerant of shade. Foresters primarily sped the process, intending to produce *the* species with high value.

To a degree, however, nature has not surrendered to a monoculture. At the time of the planting, the need to match the sources of seeds and seedlings to the particulars of where they were to grow was not widely recognized. Now we know that seeds are genetically coded to respond to countless environmental conditions. Their chromosomes help them cope. Partly for this reason, hindsight makes it no surprise that Douglas-fir have not grown uniformly well on all of the burn. On those sites of poor growth, western hemlock and redcedar are invading. They are giving the plantation token diversity in spite of human intention.

Foresters themselves are now adding diversity in the wake of a common root rot that is attacking the Douglas-fir. This is a fungus disease to which Douglas-fir is highly susceptible. It is nature's way of thinning a dense stand, an expected stage of forest development, but on the Tillamook plantation the only commercially acceptable way to handle thinning is through an interim harvest, which leaves remaining trees well spaced. Once harvesting is completed on sites where the fungus is most rampant, the Oregon Department of Forestry is fighting one facet of nature by using another: they are introducing diversity to limit spread of the disease. They are planting alder, bigleaf maple, and redcedar in two- to five-acre patches to thwart the continuing encroachment of root rot by breaking up access to its preferred host, the Douglas-fir. In nature, such a mix of species is inherent, a guard against runaway infestations.

As a result of fire some trees may be totally consumed, others scarcely harmed. Result? A "biological legacy" of dead and living material, which aids forest recovery. As ecologist Jerry Franklin says, "A little bit of chaos is a wonderful thing in a forest."

Cascades Sanctuaries

Fir-hemlock forest with bigleaf maple intermixed characterizes the green mantle of Oregon's western Cascades. Today's foresters have broadened their focus from sustaining the timber output of forests to sustaining the ecosystems that make the output possible.

On a hot summer day in 1990, I drove to a forest near the Breitenbush River in Oregon's Central Cascades hoping to see my first northern spotted owl. The forest here is classic mid-elevation old growth, all light and shadow and variegated textures with pink rhododendrons and white windflowers bordering a trail that threads among large Douglas-fir and hemlock. That year, forests like this graced posters and magazine covers nationwide as the Northwest's old-growth forests became the focus of bitter debate. As an environmental journalist, I was covering that debate in all its aspects—and the spotted owl had just become its symbol. It was now protected under the Endangered Species Act. A distinguished team headed by U.S. Forest Service biologist Jack Ward Thomas had found the owl to be "imperiled over significant portions of its range because of continuing losses of habitat from logging and natural disturbances" and concluded that "current management strategies are inadequate to ensure its viability." The team unveiled a sweeping plan: to prevent extinction, further destruction and fragmentation of habitat would be tightly restricted.

At the time, the type of forest known to harbor spotted owls was being logged at the rate of tens of thousands of acres each year. It is old forest, typified by the Breitenbush, where naturally fallen trees open holes in the lofty canopy, and where sunlight pours through, illuminating an understory of shrubs and seedlings and a forest floor heaped with decaying, moss-covered logs. The 1981 report *Ecological Characteristics of Old-Growth Douglas-Fir Forests,* written by forest ecologist Jerry Franklin and seven colleagues, could have been describing the Breitenbush.

As I walked the trail with three companions, my wish to see an owl was granted—twice over. Just as we rounded a bend, someone took my arm and pointed into the trees. A spotted owl was watching our approach with unswerving gaze. It had left its hollowed-out home in the broken top of a fir and sat on a moss-draped limb closer to the ground. Across the trail, a fuzzy white fledgling nearly as large as its parent was perched in a young hemlock that had sprouted on a wind-felled tree's decaying root. For several minutes I stood gazing at the emblematic birds with their round, expressionless black eyes. Although I had written a good deal about the owl, suddenly its plight and the whole intricate dance of life and death that plays out within and beneath the old-growth forest canopy became real to me.

By now the remarkable story of this ecosystem and how it functions is well known. Within the decaying fallen logs live complex societies of insects and other invertebrates,

which join with fungi and bacteria to slowly convert wood to soil. The process may take centuries. Meanwhile, the logs provide a home and a pathway through the undergrowth for red-backed voles and other small rodents, and these animals in turn spread the spores of fungi, which bond with the roots of enormous conifers and help to nourish them as they rise 200 feet or more, reaching for the light. The northern spotted owl shelters in this canopy, which provides protection from its chief predator, the much larger great horned owl; and from the canopy, the spotted owl swoops to the ground to snatch voles and flying squirrels and other prey.

Spotted owls are closely adapted to low- and mid-elevation old conifer forests west of the Cascade crest, so much so that the U.S. Forest Service long ago designated the birds as an "indicator species." Their presence provides a gauge for the health of the forest itself. In 1990 biologists knew of nineteen owl pairs in old-growth stands near the Breitenbush River. For me, the two owls by the trail embodied all of this. I savored the sight, then moved closer. The adult owl flew off to a higher limb, but the fledgling stayed where it was. As we hiked away, it seemed to follow us with its eyes.

The mortality rate of young dispersing owls is one of the great unanswered questions in spotted-owl research. Whether they are able to survive the flight over vast logged landscapes and find suitable habitat to propagate their kind in sufficient numbers over the next century will determine whether they escape extinction. Owl biologists like Eric Forsman, who has spent a quarter century studying the birds, believe their mortality rate is extremely high, although tracking them is admittedly an inexact science. Monitoring females is more precise than following juveniles, and a 1994 U.S. Forest Service population-modeling exercise based on ten years of survey data seems to bear out biologists' concern. It shows that breeding females have been declining throughout the northern spotted owl's range, and that their rate of decline appears to be accelerating despite logging restrictions to protect habitat. Whether the restrictions have come in time to allow owl populations to establish a new equilibrium is unknown.

In 1989 the listing of the owl by the U.S. Fish and Wildlife Service as a threatened species triggered extensive surveys on federal, state, and private land and a burst of research projects that have made the bird one of the most-studied in history. The surveys turned up owls in forests that did not meet the accepted definition of old growth. They were in younger forests of the Coast Range and California coastal redwood region, where trees grow fast and acquire some old-growth characteristics early. Owls were also found in transition zones at the eastern edge of the bird's range, where selective logging had left enough structure to support nesting and foraging. And a few were discovered in the Tillamook Burn, where surviving patches of old growth provide nesting sites and surrounding plantations supply an adequate prey base.

Such findings have increased understanding of the owl's prey and habitat requirements. They also lead to concern that even if all logging ended today, birds fledged in the twenty-first century but genetically coded by past millennia will have to survive in a landscape transformed by twentieth-century fragmentation of the forest's green cloak. Plantations will mature over decades and centuries, but extensive forest canopy as the owls have known it is unlikely regardless of how much old growth is saved. Does it matter if the owl proves unable to adapt? That is a question for philosophers and ethicists. But if it does not adapt, the intricate weave of the forest ecosystem will become more simplified, its fabric will tear in new, as-yet-unknown ways—and humans will have contributed to the further impoverishment of the earth's biological mantle.

Table Rock southeast of Portland is the Bureau of Land Management's one wilderness area in western Oregon. It is perhaps the best place in the Cascades to see forest fragmentation. A nine-square-mile mesa top forested with Douglas-fir and western hemlock, it is surrounded by a quilt of alternating mile-square patches of public and private land, mostly logged. Hiking the trail that crosses the mesa, trying to avoid the edges where clearcuts and ridge roads extend to the horizon in all directions, one has an

Facing page: *A fledgling spotted owl* (top right) *perched on a yew branch symbolizes both the threat of extinction and the forest's potential as a life-saving pharmacopeia, for yew trees are a source of the cancer-fighting drug Taxol. Spotted owls prey on red-backed voles* (bottom left), *and are themselves preyed upon by great horned owls* (top left).

inescapable sense of being on an island floating in a stark, transformed landscape.

The scene is representative. Most of Oregon's remaining old-growth forests are in the Cascades, but only a few mid-elevation watersheds—and even fewer low-elevation watersheds—remain intact. Generally speaking, those remnants are in designated wildernesses. Popular French Pete Creek, accessible year-round, was annexed to the Three Sisters Wilderness in 1978. Six years later, Congress created several "pocket wilderness areas" containing commercially valuable mid-elevation old growth; these include the Middle Santiam, the Salmon-Huckleberry, and Bull of the Woods. From a low-flying plane the reserves stand out, dark jewels set in a landscape checkered by clearcuts and second-growth plantations and laced with logging roads. Seen from the ground, these intact forests are of unsurpassed beauty and variety. In late spring and summer, their stream banks and openings are lush with trillium and wild lily-of-the-valley, wild rose and Oregon-grape, wood violet and candy flower, anemone and false Solomon's seal. In fall, fiery vine maple brighten the dark green of the conifers. In winter, snow weighs down the fan-shaped boughs of Douglas-fir and turns the landscape into a black-and-white etching.

Microclimates abound. In a rocky clearing near Waldo Lake, vine maple grows thick and impenetrable. Along the Cascades crest, or wherever a ridge creates a drier, warmer climate on a south-facing slope, Douglas-fir and western hemlock and redcedar segue into ponderosa pine. The Old Cascades east of Salem, the nonvolcanic bases of mountains that preceded today's younger Cascade Range, contain seventeen conifer species, including a spectacular grove of very old yellow-cedar. And in the Menagerie Wilderness, an upland forest of madrone and golden chinquapin seems transplanted from the Siskiyous far to the south. At higher elevations, Douglas-fir and hemlock give way gradually to noble fir with its stiff, brushlike, upturned needles, and subalpine fir with its cathedral spires. Just east of the Cascade crest, the forest changes abruptly, often within less than a mile, to ponderosa and lodgepole pine.

The need to understand the dynamics of these forests has long been recognized, though it was only in the 1980s that serious research into their workings as intricate ecosystems gained momentum. In 1948, at about the same time the Forest Service began selling timber rapidly from its lands in order to meet the postwar building boom, it also set aside 15,000 acres in the Oregon Cascades east of Eugene as an outdoor research laboratory. This is the H. J. Andrews Experimental Forest, named for a former regional forester and renowned for the research it has produced. It lies within the Willamette National Forest's Blue River Ranger District and within the drainage of the Blue River, which flows into the McKenzie River. Elevation ranges from 1,350 to 5,340 feet.

The Andrews has always been a working forest. It still is. But its role has evolved with changing views of forest management. In the 1950s it was a laboratory for trying out new ways of building roads and harvesting old growth. In the 1960s scientists there began to study how logging affects streams and soils. Then, in the 1970s, the National Science Foundation launched the International Biosphere Project to study selected ecosystems in danger of disappearing from the Earth, and the Andrews became a crucible for basic research on the old-growth forests of the westside Douglas-fir region.

Jerry Franklin, then a Forest Service ecologist, now a University of Washington professor, along with Richard Waring of Oregon State University, assembled a team of dedicated research scientists and set about studying the structure and function of these ancient forests, at the time commonly called "cellulose cemeteries." Franklin and Waring had in mind studying not just plant communities but what inhabited them, from microorganisms to insects to voles to songbirds to spotted owls—and also how all these interact. It was an unprecedented project. This type of research had never been attempted by Northwest schools of forestry, which for the most part focused on improving the efficiency and productivity of commercial forestry.

The team's work gave new meaning to the concept of long-term research. For example, Mark Harmon, now at the Oregon State University, currently directs a study of six hundred trees that were cut, labeled, and left on the ground as thirty-three-foot logs to

Old-growth Douglas-fir like that of the central Cascades, if lost, will not regrow to this stage within the lifetime of our great, great, great, great, great, great, great, great grandchildren. To responsibly steward such remaining forests, management attention now focuses on entire landscapes, regardless of present human ownership or jurisdiction. This new approach recognizes the forest as an ecological whole.

decompose. Periodically they are sampled and sections are sliced off to analyze and measure the progress of decay. Four species are involved: western redcedar, Pacific silver fir, Douglas-fir, and western hemlock. This is intended as a 200-year study, a deliberate match of the forest's own timetable.

Other research has been speedier. In 1981 the team drew up the first ecological definition of the Northwest's fast disappearing old-growth Douglas-fir–western hemlock forests. Published by the Forest Service's Pacific Northwest Forest and Range Experiment Station, this document identifies four key structural characteristics of old growth: large, live trees 175 to 750 years of age and even older; standing dead trees (snags); logs fallen to the forest floor; and logs lying in streams. From this work, most research on old-growth forests has flowed.

A snag in Ochoco National Forest stands among healthy ponderosa pine. Defoliators—insect "pests"—seldom live in snags, but carpenter ants do, and pileated woodpeckers come to the snags to feed on the ants. Their oblong cavities form nesting and roosting sites for other birds, which feed on the defoliators.

Research findings at the Andrews assumed new importance in the late 1980s as the spotted owl and its ecosystem took center stage in the intensifying old-growth debate. Looking for ways to apply the new findings, the Andrews team began working with the Blue River Ranger District to design timber sales that would ground-test experimental forest techniques. The goal was to emulate natural processes and provide a "biological legacy" for cavity-nesting birds and other species after logging. They hoped to achieve this objective by leaving varying numbers of snags, fallen logs, and even live trees on otherwise heavily logged sites. These "sloppy clearcuts" were a sharp departure from the dominant logging method of the time: cut it all, burn the slash, and apply herbicides to eliminate plants that might compete with new conifer seedlings.

Once the owl was federally listed as threatened, federal agencies quickly embraced this "new forestry" as a middle ground in the logging debate. State forestry departments and some industrial timberland owners have now followed suit. In 1991 the Willamette National Forest, the Pacific Northwest Research Station, and Oregon State University's College of Forestry formed a partnership called the Cascade Center for Ecosystem Management. Its mission is to pursue new research in the Andrews on topics that include the use of prescribed fire, landscape ecology, and restoration forestry, which embraces the concept that thinning single-species plantations and planting a variety of conifers can hasten succession to something approaching old-growth forest.

By the late 1980s, clearcut logging was well on its way to creating a two-component landscape in the Cascades: wilderness in the high country and a patchwork of forests, clearcuts, and plantations below. How will this fragmentation affect wildlife and forest health over the long term? Studies are little more than begun, but some conclusions are obvious: deer and even owls can travel considerable distances in search of food and cover. Snails and salamanders that are endemic to particular streams or springs cannot.

Eastside Remnants

The paved U.S. Forest Service road to Lookout Mountain in Central Oregon's Ochoco National Forest offers a now-rare vista: mile after mile of old-growth ponderosa pine forest. With their cinnamon-hued, puzzle-bark armor, the stands rise from south-facing slopes at low elevations, each tree solitary in its grandeur, each casting a long shadow in an open, parklike forest. These giants inspire an odd combination of serenity and awe that comes not just from their size and distinctive bark but also from their gnarled

limbs, flattened crowns, and clusters of long needles catching the sun and softly rustling in the wind.

Ponderosa pine symbolizes the old growth of Oregon's arid east side. In the moist, shaded draws of the Blue Mountains it creates green stringers amid brown hills, and at higher elevations it is the dominant species in a mixed forest that includes Douglas-fir, grand fir (here called white fir), lodgepole pine, and western larch. Pileated woodpeckers, white-headed woodpeckers, and flying squirrels readily find homes in the abundant snags, and the mosaic of forested and open areas invites mule deer and Rocky Mountain elk.

The remaining fragments of eastside old growth offer a tantalizing experience of the diversity that was commonplace as recently as the mid-1900s. For westside Oregonians who like to cross the mountains for sunshine and open vistas, the flat Metolius country on the eastern flanks of the Cascades is close, convenient, and rewarding, with the clear, spring-born Metolius River flowing beneath open stands of burnished ponderosa pine. But most of the finest old-growth pine pockets are in far more remote and rugged country. The Wenaha-Tucannon Wilderness in the Blue Mountains and the Imnaha country of Hells Canyon are lands of steep canyons and wild rivers far from population centers, and the Wallowa Mountains provide abrupt transitions between forest and high desert.

Swatches of ocean mist funneling up the Columbia River from the coast reach 300 miles inland to the northern Blue Mountains. There the moisture supports fairly lush forests, a contrast to the southern Blues where dry winds off the Snake River plain decree aridity.

Heading east in late summer, I cross Ochoco Summit in central Oregon. Quaking aspen glitter silvery green at the edge of a meadow. Nearby in a protected research area, a woodpecker hammers away at the trunk of a dead fir. A hawk circles slowly. Cattle are fenced out of this area, which allows wild grasses to grow high and conifer seedlings to sprout relatively undisturbed. These fenced, ungrazed lands are rare. On federal land east of the Cascades virtually every patch of open ground is leased for livestock grazing.

I am reminded of this again while continuing south to the pine forests flanking Lookout Mountain. Here, though the forest is set aside for backcountry recreation, cattle have altered the ecosystem just as the exclusion of fire has changed it elsewhere. They have paved the ground with their "pies," munched native grasses to bare dirt, eaten the tender shoots of cottonwood and aspen, and trampled streams. Federal law allows livestock even in designated wilderness. The forest is free of their unmistakable imprint only in research areas.

This has gone on for more than a century and has altered the forest as surely as has logging. Local ranches and outside corporations grazed hundreds of thousands of cattle and sheep in eastern Oregon beginning in the mid-1800s; particularly prized have been high—and vulnerable—summer pastures. In the late 1800s when railroads opened eastern markets, drovers herded as many as 25,000 head of cattle at a time through eastern Oregon en route to the Great Plains to fill the empty grazing niche left by the deliberate decimation of the buffalo. Effects had to be drastic. Think of the cumulative eating, defecating, drinking, and trampling.

I drive to the Blue Mountains to meet Kevin Scribner, a Walla Walla, Washington, environmental activist, who shows me the diversity of eastside forest types. We enter the Umatilla National Forest near Oregon's Meacham Summit and drive a steep logging road through dry hills until it dead-ends at a landing. Then we climb a knoll for a vista of dry hills to the east and Scribner points out green conifer stringers filling the draws. Researchers believe they got their start during periods of high moisture.

Following a ridgeline down, Scribner points out the sharp transitions between microclimates in this region of temperature extremes: ponderosa pine, with their long tap roots, grow on slopes facing south and west, where the hot afternoon sun beats down; mixed stands cover moister slopes facing north and east. Soon we stop in a cool glade with plant communities more characteristic of those found west of the Cascades: Pacific yew, wild rose, currant, nettles, trail plant, wild strawberry, beadlily, and ferns. Engelmann spruce grows here, and grand fir. This designated old-growth area, fed by a spring, is influenced by moist winds blowing off the nearby Columbia River, a phenomenon limited to the northern Blue Mountains; the southern Blues receive hot blasts from the parched Great Basin desert. We hear a pileated woodpecker and Scribner chases it through a glade. The bird prefers hollowed conifers colonized by Indian paint fungus, and they are what Scribner seeks. He wears a tee shirt urging "Save the rivers and forests, leaping with life." With so much variety in the forests around him, he has little patience for technical definitions of old growth. "We can live or die by technical definitions, when what we're really after is protecting habitat," he says.

After a century of logging and heavy grazing, eastside old-growth forests outside a few Congressionally protected preserves and research areas have been reduced to island status, creating tenuous habitat footholds for a number of species associated with old growth, and many scientists believe the composition of the forested landscape as a whole has been altered in ways that will take centuries to overcome. One example: a 1936 Forest Service inventory covering the Malheur, Ochoco, Umatilla, and Wallowa-Whitman National Forests found that about 80 percent of commercial forestland included a significant proportion of ponderosa pine stands, but no more than thirty years later only

40 percent of those same forests reported a significant proportion of ponderosa. Today many of the managed stands that replaced the pine have fallen victim to fire, drought, and epidemic insect infestations, turning large sections of the Blue Mountains and eastern Cascades into ghost forests.

How did this happen? The short answer is that twentieth-century logging of the most accessible forests is responsible; ponderosa pine terrain is gentle, and loggers took the most valuable, oldest, largest, and healthiest trees—the ponderosa—leaving Douglas-fir, grand fir, and lodgepole pine. Livestock grazing also enters in. So does fire suppression. In fact, researchers at Oregon State University have concluded that the suppression of fire in recent decades holds a major key to this forest's changing face. For centuries, small, relatively low-temperature fires swept through the eastside forests roughly every ten years. Many were started by lightning. Others were deliberately set by Native Americans who manipulated the land to produce various resources. Fires were so pervasive that they cast a blue pall over the rugged mountains of northeast Oregon and gave the Blue Mountains their English name.

Usually the Indians' deliberate burns were small, started in dense growth after fall rains had soaked soil and vegetation. In her 1995 book about ecology and culture in the Blue Mountains, *Forest Dreams, Forest Nightmares,* Nancy Langston writes that in these fir forests the goal was to create openings. This stimulated the growth of grasses and the sprouting of shrubs like huckleberry and grouseberry, browse for deer and elk and also a source of berries to fill women's baskets. After the mid-1700s, Native American fires also served a new purpose. In ponderosa-pine forests, they eliminated woody undergrowth and improved grazing for horses, which had been newly acquired through trade with tribes to the south. By opening the forests, fires also eased traveling and hunting by horseback. Indian herds had become huge by the time settlers began arriving in the mid-1800s; there were perhaps as many as a half-million horses in the Blue Mountains alone. The ecological impact must have been substantial.

Langston points out that fur trappers and settlers deplored the Indians' burns, describing them in their diaries as wantonly wasteful. They did not realize that the forests they admired had developed, in part, as a response to this deliberate burning over untold centuries. Nor did these newcomers see it as hypocrisy when they set their own fires. They burned to clear land for agriculture, and occasionally they set trees ablaze as festive bonfires, or to get rid of wasp nests. Later, sparks from steam locomotives and donkey engines used in logging quite routinely set the woods on fire, and slash burns, intended to clean up logging debris, often flared out of control. But these fires were for a *purpose.* Settlers and loggers saw themselves as working the land, making it productive,

Beetles burrowing through the bark of dead trees leave galleries filled with wood borings and fecal pellets ("frass") that attract bacteria, the spores of fungi, and other insects. All of these contribute to the breakdown of the wood, releasing nutrients back into the forest.

whereas they viewed native people not as active land stewards but as hunter-gatherers who harvested nature's bounty with little effort.

In the 1940s a new assessment of fire and a new form of forest manipulation began: fire suppression. Too much potential profit was being lost in smoke. The Forest Service came to regard fighting wildfire and saving timber as one of its highest missions. It was a mission based on societal expectations, not research. In eastside forests, a half century of fire suppression has now altered forest structure and composition by allowing shade-tolerant grand fir and Douglas-fir to invade and succeed ponderosa pine. Fire had kept the pine forest open, but the new conditions—without fire—have favored the invaders. Douglas-fir and ponderosa are both fire-resistant when mature, but they develop protective mechanisms on different timetables. Ponderosa do so at a young age, well within the usual, natural fire interval; Douglas-fir not until later. This means that Douglas-fir are naturally vulnerable to their first fire, but since frequent burns have been virtually eliminated in recent decades, the trees have proliferated. Unlike their westside counterparts, inland Douglas-fir can thrive in the understory; their seedlings and saplings tolerate shade. In fact, shade tolerance may be crucial for Douglas-fir in the dry country east of the Cascades as a means of conserving moisture.

Lodgepole pine, which requires hot fires to regenerate, has also increased as the intervals between fires have become longer. That lengthening allows fuel to accumulate, providing tinder for fires so hot they destroy soil organisms and nutrients and roar into the crowns of trees, often burning uncontrollably until extinguished by rain. Such conflagrations have become more frequent in the Northwest since the late 1980s. A result has been a widespread takeover by lodgepole, which is inherently adapted to fire.

In addition to this shift in forest composition, fire suppression has also indirectly increased insect infestations. Previously, fires had kept species like spruce budworm, Douglas-fir tussock moth, pinebark beetle, and fir engraver beetle under control by limiting the quantity, or quality, of their food. But as infestation-vulnerable Douglas-fir and grand fir began to dominate eastside forests, waves of opportunistic insects arrived to feast. Their arrival coincided with a drought cycle in the late 1980s and early 1990s, which made the trees even more vulnerable. Boyd Wickman, a retired Forest Service research entomologist who spent his entire career studying such infestations in the intermountain west, stresses that insects play a fundamental and natural role in shaping eastside forests. In one study he found that remaining trees grew faster after insects had chomped through a mixed forest, leaving skeletal stands in their wake. The survivors had been freed of competition for sun, water, and nutrients. And forest ecologists point out

that even insect-ravaged trees contribute seed for regeneration after fires and provide habitat for wildlife, among them pileated, three-toed, and black-backed woodpeckers, northern goshawks, and martens.

By the early 1990s, the changes wrought by logging and fire practices were impossible to ignore. The National Audubon Society formed a partnership with the Forest Service and enlisted volunteers to map and inventory remaining old-growth stands in eastern Oregon and eastern Washington. These activists discovered that old growth east of the Cascades was far less prevalent than anyone had suspected, and that 91 percent of

remaining ponderosa were actually in stands no larger than 100 acres. Their findings were confirmed in 1995, when U.S. Department of the Interior's National Biological Service pronounced old-growth ponderosa pine forests the most endangered of all forest ecosystems.

The decline in health of these forests also prompted federal, state, and private forest managers to begin reevaluating their stewardship. The Forest Service increased its use of controlled burns, mechanical removal of dead wood, and thinning of dense stands. Some scientists, including Wickman, began urging the protection of all remaining old-growth pine and larch as a seed source and a benchmark by which to measure the recent drastic alteration of the ecosystem.

Wildlife, of course, is affected by the forest changes. In the fall preceding the Biological Service's pronouncement, a panel of scientists convened by Congress had issued an independent report analyzing conditions in national forests east of the Oregon and Washington Cascades. Representing professional associations of ornithologists, wildlife biologists, fisheries biologists, conservation biologists, and landscape ecologists, the panel concluded that intensive logging, road building, and livestock grazing since 1950 had so fragmented the landscape that the survival of many vertebrate species is threatened.

The panel's report came in the aftermath of fires that had swept through the intermountain west the previous summer, burning millions of acres in eastern Oregon, eastern Washington, Idaho, and Montana, and posing a debate concerning recovery. The Forest Service soon announced a sweeping plan to salvage timber damaged by fire and insects, including forest tracts in areas at the time unpenetrated by roads. Congress pressed for even more extensive salvage logging, heightening tension between the long-term objective of attempting to restore the prefire character of the forests and the short-term goal of recovering valuable timber from damaged stands.

In November 1992 I toured the Starkey Experimental Forest and Range near LaGrande with Jack Ward Thomas, a Forest Service research biologist who was chosen a year later to become chief of the Forest Service. We walked through ravaged stands where sturdy green seedlings eventually would begin building a new, human-designed forest. Thomas, an elk biologist, had been instrumental in getting funding for a controversial study that used sophisticated radio telemetry to track elk and cattle fenced within 3,600 acres of mostly forested land. The goal was to study the animals' relations with habitat, humans, and each other. But in 1988, as the research design was taking shape,

The Wallowas' spectacular Imnaha Canyon breaks the green of Oregon's forested slopes, as do other canyons in the mountains northeast of La Grande. Summer wildflowers add their fleeting carpet of color to the high grasslands.

pinebark beetles and other insects began an intensive attack, and it was decided that the Starkey would also become the site of a massive salvage project designed to test various strategies for restoring diseased forests throughout the Blue Mountains.

The project became a drastic thinning operation. Loggers cut diseased and dead trees, testing various configurations of from three to fifty-three acres. In these clearings they left from four to thirty standing trees per acre, mostly resilient ponderosa pine and western larch. Over a period of two years, the process resulted in thinning half of the forest enclosed by the fence. If dead wood needed to be removed—in part to reduce fire hazard—thinning was the only available course in heavily diseased stands like this one, Thomas said. The fuel load was so great that prescribed fire was not an option. Its intense heat would destroy the organic soil layer and lead to erosion and stream temperatures lethal to fish.

Thinning is considered a "forest-health treatment." It is a practice most environmentalists tend to oppose. They see it as a subterfuge to allow the logging of healthy trees, and they question whether dead wood really needs to be eliminated. Many scientists also have reservations. They say treatments should proceed on a case-by-case basis, with restoring the forests, not capturing their economic value, as the guiding objective. Thomas pointed to the Starkey operation as a necessary, if primitive, step toward understanding how humans can begin to repair the damage they have inflicted on these forested mountains. "We were dealing with these issues one at a time without a larger view of the landscape situation over time," he said to me. "[But it] became obvious we were dealing with a long-term problem. We need a real vision of what we want this forest to be."

Related to the issue of whether and how much to thin weakened trees after an insect infestation is the issue of whether and to what extent burned trees should be salvaged after wildfires. In 1995 political pressure for stepped-up salvage of fire-damaged timber heightened in the wake of fires that had burned 4 million acres throughout the West the previous summer. Another view was presented by a panel of scientists from universities, federal and state agencies, and Indian tribes who published a paper declaring that there was no ecological need to salvage timber after wildfires. In fact, they warned that salvage logging and road building after intense fires often further damage soils and streams and slow their healing. "The problem is not that we [lack] knowledge to control all disturbances," they wrote. "The problem is we have tried to control all disturbances rather than letting them play out—[and] the forests depend on disturbances to maintain their integrity just like rivers depend on flood and droughts coming along in irregular patterns."

~

The Wenaha-Tucannon Wilderness in Oregon's extreme northeastern corner is a land of steep canyons and wild rivers far from any population center. There, from Elk Camp, I hike downhill five miles to the Wenaha River, which carved this V-shaped valley. The tall pines along the trail inspire awe not just because of their age but because of their character. They stop me again and again, drawing my gaze to gnarled and twisted limbs, to rough bark punctuated with holes bored by insects, and to gardens of needle and lichen. These mountains and canyons are heaven for big game. I see no elk this day, but their presence is palpable. In a clearing by the river, some hunter has nailed a pole between slender firs to hold a carcass.

Not far from the Wenaha-Tucannon, the Eagle Cap Wilderness of the Wallowa Mountains protects alpine forests in deep glacier-carved valleys. From East Peak high in the Wallowas, the view encompasses an abrupt transition between forest and desert. To the west and south, the Wallowas display their granite peaks, cirques, and knife-edge ridges. Beyond lie glaciers and more peaks arrayed like vertebrae in the spine of a stegosaurus. Set in a half-moon of mountains is blue Aneroid Lake. To the north are the communities of Joseph and Enterprise, and the flat irrigated farms and rolling plateaus that stretch off toward hazy canyons. To the east, one more steep forested valley lies between my camp and the highlands that roll away to the rim of Imnaha Canyon and the great Snake River divide known as Hells Canyon country. Floating above all is Idaho's Seven Devils Range. It is rugged, untracked country, unknowable even if you spent a lifetime trying. The forests here are natural fragments that took root in hospitable pockets of a dry, serrated landscape.

I drive to Buckhorn Lookout. It perches at the edge of Imnaha Canyon on the Oregon-Idaho border across a high tilt of prairie. Enormous pines grow at the lip of the chasm. Canyons upon canyons stretch off in lavender light to the north, south, and east. As light fades, a new moon emerges from behind a black cloud and a sudden wind kicks up as moonlight illuminates the pines. Beyond this last stand of great pines, the land falls away in all directions.

Washington

By Tim McNulty

Tim McNulty, poet and nature writer, formerly earned his living by working in the Northwest woods on planting and thinning, selective logging, and trail maintenance crews. As a University of Massachusetts student majoring in literature, his first view of the West was through its poets—Robinson Jeffers, Theodore Roethke, Kenneth Rexroth.

To see with his own eyes what they wrote about, Tim hiked in the Rockies, Sierra Nevada, Cascades, and Olympics, and while camped among the ancient trees of the Olympic Peninsula—"the wildest and most diverse" of the places he had visited—he decided he had found a new home. He settled near Sequim, where he lives with his wife and daughter. His most recent publications include a book of poems, *In Blue Mountain Dusk*, and *Olympic National Park: A Natural History Guide.*

Right: *Olympic National Forest*

Redcedar are long-lived (typically to 1,000 years), yet they are relatively slow growing, not particularly prolific, and vulnerable to fire because of thin bark. The widely buttressed base of aged trees often becomes hollow, but growth continues in outer tissues.

Washington

Long Island Redcedar

A flock of Canada geese fed beside a slough, and a great blue heron traced a line of marsh grass that bordered the wooded shore of the island. As our canoe struck the beach, another heron flapped off toward the open waters of the bay. Tucked into the southwest corner of Washington and surrounded on three sides by low forested hills, Willapa Bay is one of the West Coast's most important migratory and nesting bird habitats. But it was not birds my companions and I had paddled across the short passage to Long Island to see. It was a grove of ancient western redcedar, a remnant of a once-larger forest apparently little changed through the last 4,000 years.

As we entered the grove, the air grew cool and bits of filtered sunlight broke among leaves of salal and evergreen huckleberry shrubs. The old cedars rose from fluted bases like great mossy pillars leaning among long-forgotten ruins. But for the soft calling of a varied thrush, the forest stood silent. It has been almost a quarter century since I encountered my first ancient redcedar grove; still there's something in the sight of these old giants that both humbles and thrills. The cedar wore thin skeins of ragged fibrous bark, reddish chestnut near the ground and fading to silvery gray where it caught the sunlight. Hemlock saplings and huckleberry shrubs had taken root in clefts and folds of bark and atop the large burls that grew from several of the trees. Low limbs drooped from the open canopy, and high limbs reached skyward, forming the candelabra tops characteristic of old-growth cedars. These multiple broken crowns, which snap easily in high winds, combine with redcedar's relatively low stature and wide buttressed base to help the trees withstand the harsh winter storm winds characteristic of the coast. This allows cedar to reach venerable ages; some here are estimated to be a thousand years old.

Cedar of varying ages and sizes, some reaching eight to ten feet in diameter, stood widely spaced throughout the forest; smaller western hemlock grew among them. Hemlock are not as long-lived as cedar. The oldest here are estimated at 500 years; the youngest I saw was a fresh seedling sprouting on a mossy log. Together, the cedar and hemlock formed an open, multistoried canopy, interspersed with standing dead trees (called snags) and a maze of down logs: the classic structure of old-growth forests.

As is true of ancient forests everywhere, the Long Island cedar forest's multiaged nature provides varied habitats for wildlife. Biologists have documented nesting bald eagles, spotted owls, and marbled murrelets here, all of them listed by the state as species of concern, the owls and murrelets also listed by the U.S. Fish and Wildlife Service as threatened. My visit to the island grove offered none of these, but I did see the rectangular holes pileated woodpeckers had chiseled into several snags, and droppings indicated the forest provided winter habitat for elk. Fish and Wildlife Service biologist Don Williamson says that black bear, black-tailed deer, a number of small mammals, eight species of salamanders, and a dozen nesting songbirds also call the cedar forest home. Such abundance in a single remnant stand of less than 300 acres points to the tremendous productivity of low-elevation, old-growth forests.

It was this that caught the attention of Willapa National Wildlife Refuge manager Jim Hidy soon after coming to the area in 1980. He told me that something about the Long Island stand struck him as different from other coastal cedar forests he had seen: "After a single visit, I began trying to get some forest ecologists to come out and look it over." In 1984 a group that included Jerry Franklin, at the time chief plant ecologist for the U.S. Forest Service's Pacific Northwest Forest and Range Experiment Station, as well as Jan Henderson, area ecologist for the Forest Service, and Reid Schuller, ecologist for the Washington Department of Natural Resources Natural Heritage Program, toured the stand. Their visit catapulted the island cedars onto the map.

Franklin reported that this western redcedar stand constituted a community type he had never before encountered. He noted that the grove was well drained, rather than having the swampy conditions and associated vegetation expected in coastal redcedar stands. He also commented that the grove's redcedar and hemlock composition appeared stable. Henderson noted that catastrophic fires appear to have been absent for several thousand years, possibly since the end of the post–Ice Age warming of climate 5,000 years ago. It was then that sea levels stabilized and vegetation patterns similar to those of

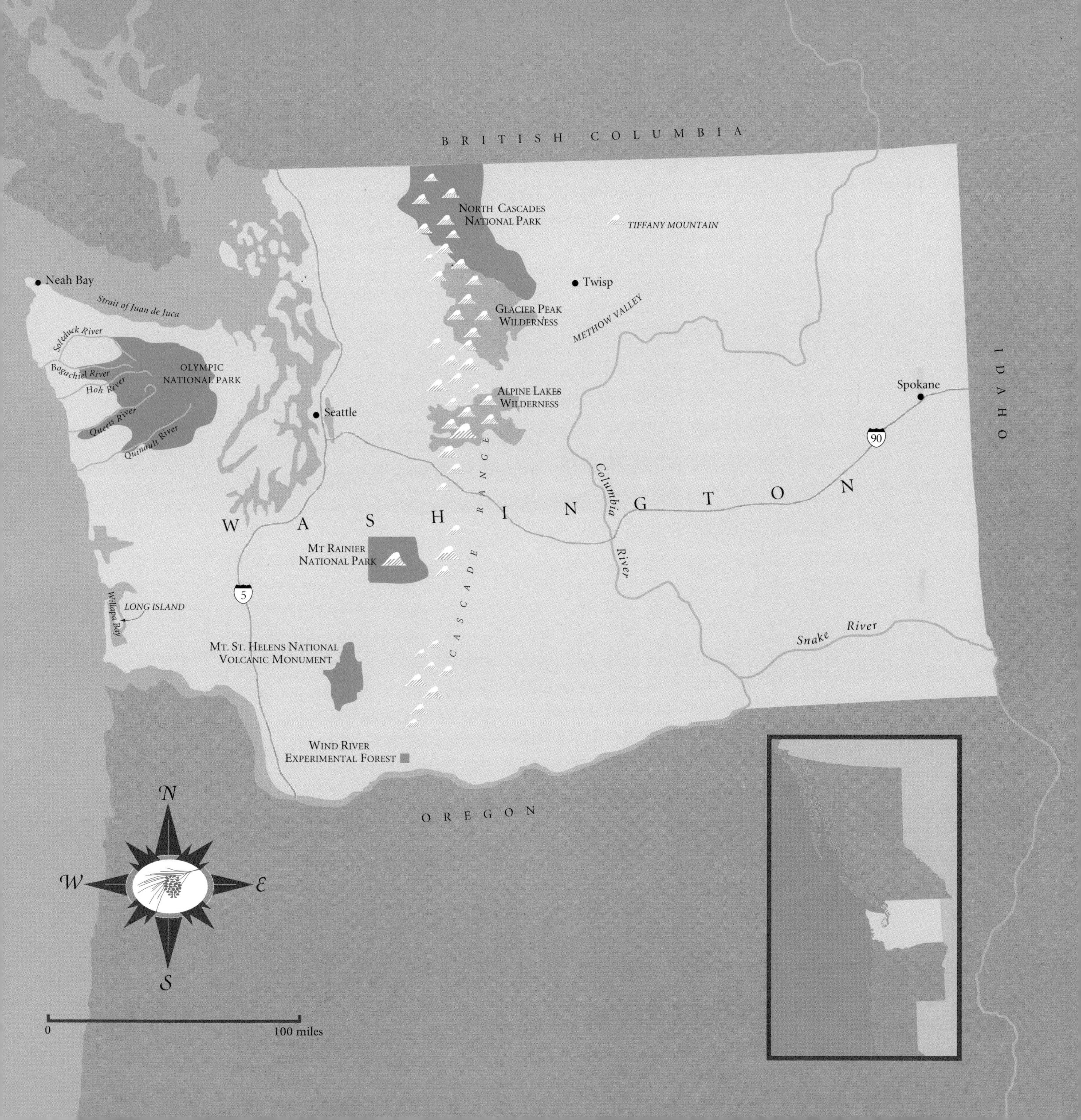

BRITISH COLUMBIA
IDAHO
OREGON
WASHINGTON
NORTH CASCADES NATIONAL PARK
TIFFANY MOUNTAIN
Neah Bay
Strait of Juan de Juca
Twisp
GLACIER PEAK WILDERNESS
METHOW VALLEY
Soleduck River
Bogachiel River
Hoh River
OLYMPIC NATIONAL PARK
Queets River
Quinault River
ALPINE LAKES WILDERNESS
Seattle
Spokane
90
Columbia River
CASCADE RANGE
MT RAINIER NATIONAL PARK
5
LONG ISLAND
Willapa Bay
Snake River
MT. ST. HELENS NATIONAL VOLCANIC MONUMENT
WIND RIVER EXPERIMENTAL FOREST
N
W
E
S
0
100 miles

today began to take hold. Before that, the pollen record and other evidence shows that species like lodgepole pine and Douglas-fir typically found on arid sites thrived along the coast. As the climate cooled and became wetter, western hemlock, Sitka spruce, and western redcedar gradually replaced them. Even now, however, small patches of lodgepole, or shore pine, as it is called along the coast, still linger, an echo of that drier time. Henderson estimates that around 4,000 years ago—about the time the Egyptian pyramids were being built—conditions were right for the Long Island cedar forest to begin its long journey toward what used to be thought of as climax, a stage of growth seldom actually achieved but apparently one that exists on Long Island.

The term "climax" has become a slippery one among forest ecologists. It describes a hypothetical, self-sustaining state, the end point of ecological succession. A climax forest community supposedly maintains its mix of species and ages indefinitely or until some disturbance such as a windstorm, wildfire, or outbreak of disease cycles it back into an earlier stage of development. Actually, disturbance almost always outpaces this theoretical "ultimate" situation. Nearly all Pacific Northwest forests, including most ancient forests, are patchy with various stages of succession. The variables that forestall climax are just too many. Even aside from windstorms, wildfires, disease, and human timber consumption and land clearing for farms and cities, disturbance can come from volcanic eruptions, mudflows, glacier advances, inundation by glacial meltwaters, fluctuating sea

Disturbance plus healing contribute to forest dynamics. Winter rains often swell the Olympic Peninsula's Hoh River (far left), *causing it to flood. Occasionally mudflows engendered by water bursting from under a glacier veneer the land, as at Mount Rainier in 1987* (left). *In both circumstances, individual plants are ripped out; soils are covered by fresh silt and rock; life resumes.*

levels, rapidly shifting climate, and a host of other factors. For a forest in the heavily logged coastal lowlands of southwest Washington to escape all this strikes me as unfathomable. But the Long Island cedar stand persists: new trees grow up through the centuries; old trees die and house generations of snag-nesters or fall to the forest floor and host other vertebrates and invertebrates as well as shaggy carpets of seedlings. Through millennia, the grove has deepened in complexity and beauty.

Western redcedar was the cornerstone of the native cultures of the Northwest coast. Nowhere in the Americas is a tree more closely linked with a material culture. Lightweight, resistant to rot, easily worked and split into planks, redcedar was the wood of choice. Coast people built their great communal longhouses of cedar posts and planks, and they shaped and hollowed cedar trunks into seagoing canoes. Women shredded and wove the pliable inner bark of cedar into mats, blankets, capes, skirts, and hats; they even wrapped and diapered their newborn babies in soft-shredded bark.

The extent of human involvement with redcedar came vividly to life a few years ago as archaeologists investigated a village site on Washington's northwestern coast. About 450 years ago, well before Europeans first reached the Northwest coast, a mudflow buried several houses at Ozette, a village just south of what is now the Makah Indian Reservation. The wet clay of the slide sealed and preserved thousands of wooden artifacts that otherwise would have decayed. An excavation, led by Richard Daugherty of

Washington State University, yielded a wealth of objects from the daily lives of the past. Among them were cedar-bark mats and baskets, cedar house planks and uprights, boxes, and tool handles. Some items were decorated with exquisitely carved designs; some probably figured in ceremonies associated with whale hunts, seal hunts, or fishing. Now housed at the Makah Museum in Neah Bay, the Ozette artifacts offer a window onto life at the edge of forest and sea, and a testament to redcedar's place at the heart of Northwest coast culture.

The Olympic Rain Forest

When fires flickering within cedar longhouses were the only lights on the Northwest coast, dense forests dominated by Sitka spruce flanked beaches from northern California to Southeast Alaska. In the protected lowland valleys of Washington's Olympic Peninsula this forest reached its greatest development. Olympic National Park preserves part of that legacy, and millions of visitors are drawn there each year. They enter the valleys of the Bogachiel, Hoh, Queets, and Quinault Rivers to see Sitka spruce and Douglas-fir that tower as much as 300 feet above the forest floor. They walk beside glacier-fed rivers, breeding grounds for runs of wild salmon and steelhead. And with luck, they glimpse the Roosevelt elk that browse the ferns, shrubs, and grasses of the valley floors.

The haunting face of an owl-like ceremonial club found at Ozette dates to about 450 years ago, and evidence of people in the Northwest reaches back almost 12,000 years. The odds of artifacts being both preserved and discovered are very slight. At Ozette wood and fiber remained intact because they were constantly waterlogged, a protection against decay.

Spring has always been my favorite time in the rain forest. New leaves have yet to obscure the luxuriant growth of mosses, liverworts, and lichens that drapes the low, spreading branches of maples and alders and swathes the limbs of the high forest canopy with scarves of translucent green. Sunlight filters down to illumine the pale blossoms of spring beauty, foamflower, and yellow wood violet. Large mossy logs lie strewn about, host to vigorous spruce and hemlock seedlings. Logs musky with decay support trees up to several feet in diameter; the stilted roots of many rain forest trees reveal their origins on such nurse logs. When I stop to follow the great bole of spruce or Douglas-fir from forest floor to the upper canopy, my perception changes and routine senses of scale and proportion slip away. In the rain forest, thoughts intuitively yield to a slower and grander pace.

A number of factors led to the current development of this impressive forest community, among them the orientation of the rain forest valleys to the sea. During the two million years of the Pleistocene epoch, which ended about 10,000 years ago, alpine glaciers advanced repeatedly down these valleys. The ice widened and flattened the bottomlands and covered them with glacial rubble. The broad, low valleys that remain as the glaciers' signature are aligned with prevailing winds and act as conduits for coastal weather. Moist marine air floods them, rises against surrounding mountains and ridges, and cools. This causes precipitation to increase dramatically with distance inland. At Hoh Ranger Station it is measured at 140 inches per year; eleven miles farther up-valley at Mount Olympus it often tops 200 inches. Rains soak the forest most of the year. Even during the brief summer drought, fog drifts in from the coast, sifts through the forest canopy, condenses, and falls as "tree drip." This moisture may be critical in reducing the frequency and intensity of fires and in maintaining an essentially coastal type of forest well inland.

The open nature of the forest floor may depend in large part on continual browsing by Roosevelt elk, a grooming that creates meadowlike carpets of oxalis and sword fern and seems to influence forest composition. To study this relationship, researchers in the 1930s and 1950s built a series of exclosures to keep elk and deer out of quarter-acre plots and thereby permit the monitoring of changes in vegetation that was no longer browsed or trampled. In 1980 two more plots were added, each measuring two and one-half acres. In the early 1990s Andrea Woodward of the University of Washington and scientists from Olympic National Park checked all the plots they could find, some of which had by then been free of grazing for more than half a century. Their report indicates a marked *decrease* in grasses and forbs within the exclosures and a dramatic *increase* in the size and density of shrubs such as salmonberry and huckleberry, and of small trees such as vine maple. Herd size in these valleys has remained relatively stable since the 1950s, a situation that is rare among ungulate herds and may reflect a finely tuned balance between browse and browsers. Yet what is the record of a half century compared with the age of old-growth forest—and with the 5,000 years that the climate has been much the same as now? The limited scale of our human lifetimes skews perception.

It is the size of the trees and the lushness of the floor that rivets most visitors' attention, but scientists are increasingly drawn to the forest canopy. That unmapped realm of arboreal mosses, ferns, liverworts, and lichens is as rich and diverse as the ground is, and it also hosts a surprising array of invertebrates, birds, and small mammals. Furthermore, it is home to little-known microorganisms that perform a host of functions apparently essential to ecosystem health. Conifer needles actually harbor microscopic fungi, algae, bacteria, and yeasts both inside and on their surfaces. New research suggests that some of these organisms act directly as natural insecticides, controlling minute, harmful herbivores, and that others serve as a food base for predatory insects and spiders, which further guard against the herbivore attacks.

Temperate rain forest totals only about seventy-five million acres worldwide, two-thirds of it along North America's Northwest coast. Chile, New Zealand, and southern Australia are the other major locations of temperate rain forest. Tropical rain forest is far more extensive; it blankets most of the land between the latitudes of twenty degrees north and south.

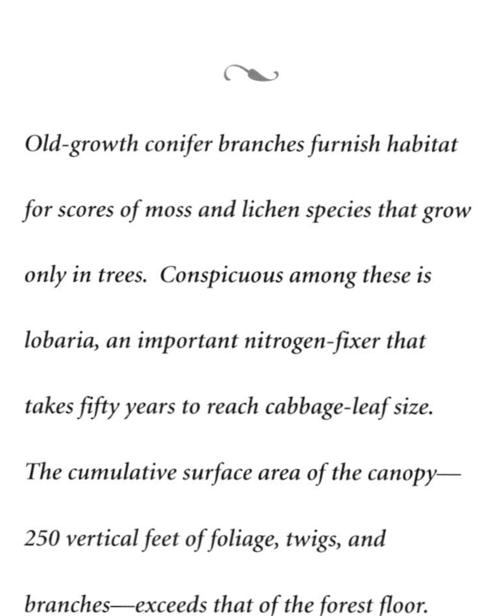

Old-growth conifer branches furnish habitat for scores of moss and lichen species that grow only in trees. Conspicuous among these is lobaria, an important nitrogen-fixer that takes fifty years to reach cabbage-leaf size. The cumulative surface area of the canopy—250 vertical feet of foliage, twigs, and branches—exceeds that of the forest floor.

Most canopy interactions are impossible for us to see, but lichens are different. A mutually dependent combination of fungi, algae, and sometimes cyanobacteria, lichens are big enough to be conspicuous, especially when blown down by winter storms. Once on the forest floor they provide an important winter food for elk. Some lichens, like lobaria, are able to assimilate nitrogen from air and rainfall, making it available to the biological processes of the forest. Within the trees they are part of the massive epiphyte gardens that form on limbs, producing mats and drapes as much as six inches thick and three to five feet long.

Epiphytes grow without parasitizing their hosts (*epiphyte* means "on plants"). In the rain forest these aerial gardens attain their most elaborate development on the wide-spreading limbs of bigleaf maples. There, lush draperies of moss and selaginella club-moss, wisps of lichen, and lacy clusters of licorice fern sway overhead. It was long believed that the maples and other trees serve only as toeholds for these communities, but that changed in the 1980s when Nalini Nadkarni, then a graduate student at the University of Washington, used climbing ropes to ascend into the forest canopy of the Hoh Valley. She made a startling discovery. Within the epiphyte mats of bigleaf maple she found an organic soil developing out of decaying plants, dust, and detritus from countless invertebrates. And beneath these mats were tree roots. They sprouted from limbs, wound along them, and tapped nutrients from the arboreal soils. Further research showed that, to a lesser extent, vine maple, alder, and cottonwood also penetrate their epiphyte mats with roots, all as anatomically complete and functional in absorbing water and nutrients as their counterparts in forest soils. Nadkarni estimated the dry weight of the aboveground epiphytes on rain forest maple at four times the weight of the trees' foliage.

To hear more about the intricacies of this newfound realm, I attended a field seminar in canopy ecology sponsored by the Olympic Park Institute at Lake Crescent, on the Olympic Peninsula. The instructors were David Shaw, an ecologist who manages the University of Washington's new canopy research facility, Nalini Nadkarni, and her student Joel Clement. About a dozen of us sat in the shadow of a tall Douglas-fir beside the blue-green waters of the lake while Shaw unfolded carefully labeled packets of canopy epiphytes. He was explaining an epiphyte classification system developed by Bruce McCune of Oregon State University based on epiphytes' ecological function within the overall forest ecosystem. A bonus is that the system also helps in understanding the relation between epiphytes and forest age.

First out of the packets were bryophytes. These are the common mosses, leafy liverworts, and clubmosses such as selaginella, which cloak the limbs of bigleaf maple in the rain forest; also the familiar shag moss that covers vine maple. Bryophyte communities are generally shade-loving and tend to be dominated by just a few species. Next were cyanolichens such as lobaria. Champion nutrient providers, the lobaria assimilate nitrogen from the air and add it to the soil as fertilizer at a rate of two to six pounds per acre per year, as measured in one Oregon forest, a large percentage of the forest's total nitrogen needs. A third group of packets held alectorioid or pendulous lichens, such as old man's beard. These usually grow on the trunks of trees. They are an important seasonal food for flying squirrels as well as deer and other ungulates—particularly at middle elevations where lobaria drops out of the canopy community. Alectorioids command special research attention; they are extremely sensitive to air pollution and therefore serve as excellent indicators of air quality.

Then Shaw opened his final packets, which held a group of lichens still poorly understood. These are mostly tube lichens, lumped by McCune as "others." They do not assimilate nitrogen from the air or provide food for wildlife. They are typically the first to appear in young forest stands. Little else is known about them.

Each functional group appears at a particular stage of forest development. In the Oregon Cascades where McCune worked, young fifty-year-old stands are dominated almost exclusively by tube lichens. Alectorioid lichens begin to appear as forest stands age. Bryophytes and cyanolichens wait until stands reach about 200 years of age, and not until 400 years do they dominate the canopy. Each group also favors its own part of the canopy: tube lichens the very tops of trees, bryophytes the lower tiers, cyanolichens and alectoriods the middle range, an overall geometry that provides nearly 200 vertical feet of toehold. This stratification creates distinctive microenvironments that help shape a dynamic ecosystem. Furthermore, precipitation drips from one canopy layer to the next, creating a "throughfall" that is chemically altered by contact with the abundance of epiphytes, microorganisms, and detritus. The result is a complex brew rich in nitrogen, phosphorous, and other nutrients.

Understanding canopy functions should prove valuable to foresters who wish to remain sensitive to ecological processes while also managing timber stands commercially. Epiphytes obviously are a fundamental part of the forest system, and, as knowledge of their various roles increases, that insight may prove to have practical applications; aspects of natural canopy interactions may be purposely replicated in forests managed by humans.

Additional studies will reveal assorted other secrets hidden in the rain-forest canopy. Already Nadkarni and her students from The Evergreen State College in Olympia have found that arthropod communities in Quinault Valley trees—mites, which feed on fungi and detritus, springtails, which are plant feeders, spiders, millipedes, and such—are as numerous in the canopy as in the soil (though the forest floor holds a greater diversity of species).

But David Shaw stresses that canopy science is still in its childhood. Worldwide, forest canopies have been called "the last biological frontier." Who can say what mysteries remain to be unlocked in the lowland rain forest of the Olympic Peninsula, renowned both for sheer biomass and for complexity? Answers seemed imminent in the early 1990s, when Shaw served as site director for the establishment of a crane that would lift researchers 240 feet into the treetops. Similar to a crane operated by the Smithsonian Tropical Research Institute in Panama, the University of Washington's crane is designed to provide scientific access to 5.6 acres of old-growth canopy; its boom is 279 feet long. In the spring of 1994, however, the U.S. Forest Service buckled under pressure from outspoken opponents of old-growth research in the timber-based communities of the western Peninsula. They blocked plans for placing the canopy research crane in Olympic National Forest. It is located instead near the Columbia Gorge, in the Wind River Experimental Forest, which is part of Gifford Pinchot National Forest. There the crane itself and plans for its accompanying research facility attract international attention. Residents of hard-strapped communities on the western Olympic Peninsula may well discover they have shot themselves in the foot.

Viewed from the canopy research crane at Wind River, Douglas-fir (top and right), *redcedar* (lower center), *and western hemlock* (lower left), *stand with branches fanned to the light. The irregular geometry of the canopy creates openness and affects air circulation. Many of the trees have dead tops. One—a 180-foot Douglas-fir—hosts a bushy, nutrient-starved hemlock seedling.*

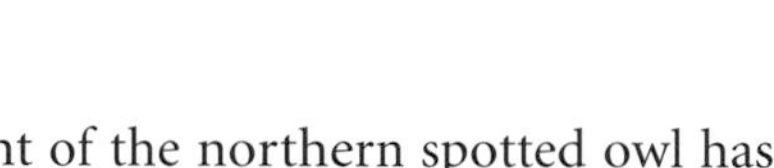

The plight of the northern spotted owl has drawn recent public attention, but the bird's place within the ecosystem is part of a linkage that extends from the owls, to the flying squirrels they feed on, to truffles, to tree roots. It is a complex of connections largely invisible to our eyes and only recently detected.

Spotted owls on the Olympic Peninsula are close to the northern extent of their range. They are an isolated population separated from others by forest fragmentation. Still, there are an estimated two to three hundred pairs here, mostly within Olympic National Park. The birds are linked to old growth because of its large, broken-top snags, which they use for nest sites; its open, multistoried canopy, which they use for hunting and roosting; and its supply of small mammals, which they rely on for food.

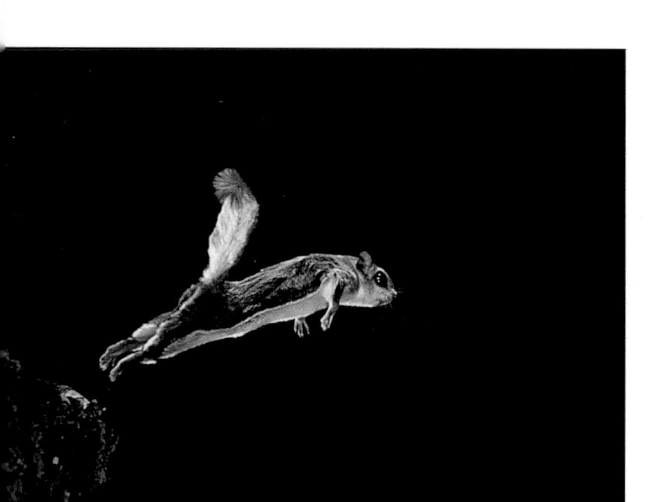

As a flying squirrel begins its glide, a loose flap of furry skin stretches taut from its front to its hind legs. Maneuvering is achieved by raising and lowering the front legs, and the long, flattened tail acts as a stablizer. From a height of 60 feet, a squirrel can parachute 150 feet.

Flying squirrels are the owl's preferred prey species. Squirrels spend much of their time in the canopy feeding on lichens, but they also have a taste for truffles. Indeed, truffles make up as much as 90 percent of their diet. Truffles are the spore-laden "fruit" of below-ground fungi, many of which form mycorrhizal relationships with tree roots. *Mycorrhiza* means literally "fungus-root" and refers to a symbiotic association in which threadlike mycelia grow on or within root tissues and crucially enhance their absorption of moisture and nutrients from the soil. The fungi stimulate root growth and also deter soil-borne pathogens and root-grazers. In turn, trees pass photosynthates—sugars and amino acids—through their roots to the fungi, which cannot produce food for themselves. These elaborate underground networks even transfer carbon and nutrients from tree to tree, enabling shade-bound trees to benefit from the photosynthesis of neighbors that have succeeded in reaching the canopy above. Some 60 percent of the carbon produced by photosynthesis actually goes to roots and mycorrhizae within the soil. This symbiosis begs the question of individual organisms living independently and points instead to the unity of the forest ecosystem.

Truffles send out a range of odors that are seemingly irresistible to the small mammals that dig and eat them, and born of this attraction is a discovery by fungi specialist James Trappe, at the time with the U.S. Forest Service's experiment station at Oregon State University, and zoologist Chris Maser, at the time with the Bureau of Land Management. While working at the H. J. Andrews Experimental Forest in Oregon, the two found that spores from truffles readily survive in the intestinal tracts of flying squirrels and other small mammals such as mice and voles. In fact, the spores become concentrated and chemically altered, and thereby increase the likelihood that fecal pellets will inoculate tree roots. It is while digging down and poking their heads into the soil to garner truffles that flying squirrels make easy targets for spotted owls and other predators. Fortunately for the forest, truffles seem to be worth the risk.

The more scientists look—high in the canopy, below the surface of the ground, or on the forest floor—the richer becomes our awareness of complex webs. Large logs lying prostrate and trees that remain standing as snags long after they die are prime examples. Snags provide nesting sites for woodpeckers, owls, raptors, Vaux's swifts, bats, and numerous small mammals—more than fifty species in Washington's westside forests. And that is just a beginning. When these trees, or others still alive, are blown down by storms, toppled by disease, or undercut by rivers, they continue to serve a variety of ecological functions. One is structural, as habitat for shrews, mice, salamanders, and countless beetles and other wood-borers. Another has to do with recycling. Insects feed first on the cambium and sapwood of logs, then on the interior wood. As they bore, they introduce fungal spores and bacteria, decomposers that lie largely hidden from human sight. It is the surface of down logs that give us the most vivid images of life drawing upon death.

Few seedlings can take root and successfully compete for nutrients among the already established plants that form a thick carpet covering the rain-forest floor. But the relatively warm, textured surfaces of the fallen trees provide excellent nurseries, particularly after an initial growth of moss has formed a damp bed for seeds. As many as fifty seedlings may crowd a single square foot of a nurse log, but only one among thousands will develop into a mature tree. Competition, sloughing bark, and toppling by wind and snow take their toll. Seedling roots may find insect borings and tap the fungi-rich decomposing interior of the log, but eventually they must make contact with soil if the tree is to survive. Enough succeed that nurse-log generated trees are common throughout the rain forest. Colonnades of huge trees with stilted roots still grasp the hollow left by a long-vanished log, bearing eloquent witness to the eternal cycle.

In the rain- and fog-dampened forests of the Washington coast, wildfire hardly qualifies as a frequent disturbance—but wind is another matter. It creates snags by snapping trees, nurse logs by uprooting trees. Windstorms blast the coast at hurricane force an average of once every two decades, and the wind's legacy is apparent everywhere in coastal forests. I felt its fury in the winter of 1979. My tree-planting crew was camped south of the Hoh Valley near the coast when a storm hit. All through the night

Colonnades of spruce 200 to 300 years old are a hallmark of coastal forest and an indirect product of wind, which probably blew down the nurse log that gave them their starts in life.

winds gusting to hurricane force ripped the forest and howled through the logged-off clearing. Trailers rocked, and tents were swept away in tatters. Remarkably, no one was injured, but a tree smashed through one van, and we lost our kitchen shelter and half our supplies. It took two days to cut through blown-down timber and reach the highway, a distance of less than a dozen miles.

Earlier, the 1962 Columbus Day storm had reached wind speeds of 170 miles per hour and felled an estimated eleven billion board feet of forest in Washington and Oregon. Even that was not the "big one" for residents of the Peninsula's west end. That honor goes to the fabled blow in January 1921, which was concentrated on the Washington coast and therefore particularly devastated the Peninsula. Winds rifled a swath thirty miles wide and took down an estimated seven to eight billion board feet between the Columbia River and the north end of Vancouver Island. Some of this downed timber was salvaged, but much of it was left where it lay. Now, more than a half century later, biologists surveying the natural stands that grew up within the blowdown have made several interesting discoveries, some that cycle back to spotted owls and flying squirrels, some that are on the verge of becoming a part of commercial forestry.

In the area devastated by the '21 Blow, there are spotted owls—although these stands, naturally regenerated, are not even a century old. There also are about twice as many flying squirrels as in nearby managed forest stands of equal age. Seeking explanation of these coinciding populations, researchers noticed that ample numbers of large standing snags and down logs—which are typical of old-growth forests—are also present in '21 Blow stands. These structural elements provide habitat and accommodate fundamental ecological processes. Ecologists term this the biological legacy of the forest, present to varying degrees in all natural forests but usually missing in intensively managed stands. The presence of old-growth-like structure in the blowdown area attracted a team led by Jerry Franklin, by then Bloedel Professor of Ecosystem Analysis at the University of Washington. They investigated a new approach to commercial forestry, one capable of producing timber while maintaining enough live trees, snags, down logs, and such to preserve this legacy and therefore the forest's future.

In an attempt to incorporate some of these new ideas into current timber management, Washington's Department of Natural Resources created the Olympic Experimental State Forest, which embraces over a quarter-million acres of state-managed forest on the western Olympic Peninsula. Early projects include a comparative study to see what strategies will best protect green trees left standing within logging units and some of the first comprehensive landscape planning for western Peninsula watersheds.

Wind also directly accounts for broken tree trunks, termed "snags." If global climate warms, wind patterns may be affected and forest blowdowns therefore substantially altered. This could change various aspects of today's forest ecosystem.

Forest Recovery at Mount Saint Helens

Disturbances to Washington's forests can take dramatic forms. That became shockingly clear the morning of May 18, 1980, when a magnitude 5.2 earthquake triggered an explosive eruption of the state's youngest and most active volcano. After 123 years of slumber, Mount Saint Helens awoke violently as its northern flank collapsed, burying Spirit Lake and the North Fork of the Toutle River beneath hundreds of feet of avalanche debris. The volcanic blast that followed swept across the forest with devastating, superheated winds of 400 to 600 degrees Fahrenheit moving at speeds greater than 600 miles per hour. Foliage vaporized, pitch boiled, and trees slammed to the ground. Avalanche debris filled twenty square miles of the upper North Fork Toutle River Valley to an average depth of 150 feet, and floods of mud and rocks hurtled into the lower valley. Streams to the south and east, swollen with snow- and ice-melt toppled trees and took out bridges. Ash and tephra rained down over the ravaged landscape. Eventually the ash cloud circled the globe.

When the air at Saint Helens cleared, observers found some 230 square miles of forest within a 15-mile radius north of the peak had been leveled. Around this blast zone, a "ghost forest" of scorched, but still standing, dead trees marked the edge of the devastation. Beyond them the living forest stood in sharp contrast, ash-laden but otherwise unaffected.

This was not the first destructive episode for the mountain; explosive eruptions in 1480 and 1800 also destroyed great swaths of forest, though not nearly as large as those of the 1980 blast. This new eruption ranks as catastrophic. Even so, in the weeks immediately following it, scientists entering the blast zone were astonished to find that some plants and animals had survived. At first small mammals were most noticeable, especially species that live or hibernate below ground or under logs. Among these, pocket gophers quickly attained hero status: their tunneling pushed organic soil up over the ash layer bringing along seeds, which readily sprouted. The gophers' excavation piles also provided nutrient-rich beds for seeds that blew in from beyond the devastated area. Bulbs and tubers cached below ground in burrows sprouted, too—another boon.

Many deer mice and jumping mice, as well as gophers, escaped harm during the eruption, safely tucked within the ground. Emerging onto the stricken landscape, they thrived and aided the blast zone's still-tentative return to life. Large numbers of insects and spiders also survived—under ground, in rock crevices, on rotting logs, under water. Ants and beetles provided ample prey for wolf spiders that ballooned in on the winds and for tiger beetles blown in from afar. Centipedes and millipedes fed well on decaying plant matter. No birds or large mammals directly survived the blast, but their kind soon returned. Among birds, the first to recolonize were those able to exploit the insect base—for example, dark-eyed juncos and northern flickers. Ragged coyotes were sighted within two weeks of the blast, making do on a scant diet of small mammals and insects—even swallowing the ash itself, perhaps along with the insects. Elk soon followed.

Green plants, too, quickly appeared. Researchers such as Peter Frenzen, now staff biologist at Mount Saint Helens National Volcanic Monument, found the forest's biotic legacy initiating its recovery in many ways. Seeds in organic soils that clung to upturned rootwads sprouted. So did those in soil newly exposed along streams. This streamside vegetation would later prove crucial as travel corridors for plants and animals moving into the blast zone. Easily dispersed plants like fireweed, pearly everlasting, and thistle soon seeded in via the wind from clearcut areas. And snow melting from north-facing slopes and frozen high-country lakes released a full complement of life-forms that had been protected by its insulating blanket at the time of the blast. Montane tree saplings like Pacific silver fir and mountain hemlock sprang up into the sunlight, and huckleberry offered fresh green leaves.

Fall and winter storms brought more surprises. Rains eroded gullies into steep ash-covered hillsides, hitting streams that had been recovering with a second glut of ash and silt. But by the following spring—a year after the eruption—these newly formed gullies turned green as preeruption plant survivors reemerged. By providing shade and organic litter and helping stabilize steep slopes, the gully plants became conspicuous players in the increasing momentum of overall recovery.

Two years after the eruption Congress set aside 110,000 acres as Mount Saint Helens National Volcanic Monument. Its purpose is to let recovery follow its natural course without human intervention, facilitate research, and help the public understand and enjoy the area as the process of recovery unfolds.

Before establishment of the monument—in the years immediately following the blast—the Forest Service and timber companies had salvaged some of the 1.6 billion board feet of down logs, then had hand planted seedlings. Ours was one of the reforestation crews planting some of the sites. Thinking back to that time, I remember the drab, gray, seemingly lifeless blast zone, a marked contrast from the lush old-growth forests of the nearby Green, Lewis, and Cispus river valleys I had explored. And I remember the dust and rumble of as many as 250 loaded log trucks per day.

More than a decade has passed since then, so recently I decided to go to Saint Helens for another look. Soon after arriving at Meta Lake, while I stood absorbed in watching a pair of mountain bluebirds nesting in a broken snag, Peter Frenzen drove up. For him, as for many researchers, Mount Saint Helens' eruption gave a new direction to fieldwork. At the time, Frenzen had been just finishing a study at Mount Rainier of how forest has reclaimed an area buried in 1947 by the Kautz Creek mudflow. "I literally drove down to Saint Helens from Rainier, climbed into a helicopter with Jerry Franklin, and began work here," he told me.

At Kautz Creek he had identified key factors influencing recovery, and he soon found those same mechanisms coming into play at Saint Helens. For instance, at Kautz Creek he noticed that the rate of plant recovery was greatest in areas with standing snags and down logs—a legacy of the original forest—and that proximity to remaining old growth clearly played a role. He found this to be true at Saint Helens as well, but the story was much more complex. That interested me. Indeed, the whole process fascinated me. On the way to Meta Lake I had driven through the managed stands where timber had been salvaged and the land replanted. They now supported dense, even-aged Douglas-fir. We had done our work well. By contrast the recovering forests inside the monument seemed sparse in places, nonexistent in others.

What Frenzen and his fellow researchers discovered is that, left alone, different sites recover in different ways and at different rates. On blowdown sites where the understory was protected by deep snow, the lower strata of the forest community remained intact; tree seedlings, ferns, and low shrubs like huckleberry and dwarf bramble survived. Of these, Pacific silver fir responded particularly well to the blast's elimination of the overstory. In places now—a decade after the eruption—it reaches heights of ten feet or more. Frenzen feels quite sure that without the seedlings' survival within the snow, the fir, and also western and mountain hemlock, would still be absent from the blast zone; these species could not sprout in the baked, posteruption earth but grew readily with the snow melted and the competition for light removed.

In contrast, in the scorched, standing forests where snow was not present, vine maple that resprouted from roots is the dominant tree species and a different complex of ground-cover plants has come up through the ash layers, for example, pyrola, pipsissewa, and bunchberry dogwood. By 1987—seven years after the eruption—Frenzen had identified 117 plant species in the blast zone. Most common were fireweed and pearly everlasting. Growing well on the disturbed soil of clearcuts, these plants survived the ashfall and sent their seeds drifting onto the stark, denuded lands.

Fireweed blossoms quickly formed a contrast to the stark posteruption devastation at Mount Saint Helens. Its dandelion-like seeds ride air currents and readily germinate in disturbed soil, an ideal combination for a pioneering species. Silver fir seedlings protected from the 1980 eruption by snow now also thrive as oncoming forest among skeletons of yesterday's forest.

"It's interesting," Frenzen told me. "The area showing the best primary recovery is the pumice plain—the place that experienced the most total destruction." There, hot pumice and gas spilling down the volcano's north side reached temperatures as high as 1,200 degrees Fahrenheit. Unlike in the rest of the blast zone, here nothing survived—above or below ground. All organic matter was destroyed. Yet now Frenzen's plots show the healthy beginnings of a new forest: sapling Douglas-fir, noble fir, silver fir, western white pine, lodgepole pine, western hemlock, mountain hemlock, and hardwoods like birch, cottonwood, and alder. Seed for some of these had to come from trees more than five miles away. Once it arrived, the ability of the porous pumice soil to hold both seeds and moisture probably explains the rapidity of germination and regrowth here in contrast with that on the scorched surfaces elsewhere. Also recovering well are the mudflows of the lower Toutle and other streams, which are similar to the Kautz mudflow Frenzen had studied at Mount Rainier.

"We're learning things here at Saint Helens that are allowing us to go back into the replanted stands outside the monument and try different planting regimes, different mixes of species and spacing," Frenzen said. "In fact, when you combine the managed parts of the blast area with the diversity of natural sites and responses within the monument, this place offers an incredible long-term laboratory for studying various approaches to managing and restoring forest ecosystems elsewhere."

Fireweed blossoms quickly formed a contrast to the stark posteruption devastation at Mount Saint Helens. Its dandelion-like seeds ride air currents and readily germinate in disturbed soil, an ideal combination for a pioneering species. Silver fir seedlings protected from the 1980 eruption by snow now also thrive as oncoming forest among skeletons of yesterday's forest.

There was one part of the forest ecosystem, quite nearby, that I wanted to visit, to remind myself what the land had been like—and eventually will be again. The Green River heads in high lake country just north of Mount Saint Helens and flows west to join the Toutle. Protected by peaks and high ridges, a four-mile stretch along the Green survived the eruption unscathed, though the forest of its headwaters and much of that farther downriver were destroyed.

Hiking in from the blast zone, I felt the air grow cool. The cheerful song of a winter wren freshened the morning. Hemlock and silver fir stood interspersed with venerable Douglas-fir four to six feet in diameter. The trail wound through a soft green carpet of oxalis, foamflower, and oak fern. Large snags stood among the trees, and mossy logs littered the ground. Farther down the valley, the stands on river benches graded into almost pure Douglas-fir of immense size. This was the preeruption forest I remembered in the valleys surrounding Mount Saint Helens and it was reassuring to walk in its shade. I thanked the ridges behind me for this blast-shadow remnant of forest.

Hiking out that evening I stopped where the forest opened onto snags marking the edge of the blast zone. Here was contrast. Between forest margin and devastation, a thick growth of young Douglas-fir and hemlock, alder and vine maple had sprung up. Battered, blast-killed trees stood like aging sentries while the new forest slowly gathered itself beneath them. Two varied thrushes called back and forth from within the woods and a pair of common mergansers preened on a sandbar in the river. It will take centuries for this seedling forest to approach the condition of the old-growth stand I had just left, but centuries are as minutes or seconds to nature's process, and the forest has been through all this before.

West Slope of the Northern Cascades

Stretching north from Mount Rainier to the Canadian border, the Mount Baker–Snoqualmie National Forest encompasses more than 1.7 million acres along the west slope of the Cascade Range. Nearly a dozen major rivers drain these rain- and snow-saturated forests; their valleys and slopes harbor some of the oldest and richest forests anywhere in the Cascades. A hundred stands spring to mind when I think of them—not the roadside forests that wink past alluringly, but forests I have traveled on foot. In my mind, I see the huge western redcedar along the lower North Fork Sauk trail and the giant Douglas-fir along the North Fork Skykomish on the Blanca Lake trail. I remember the rainy fall afternoon I first visited the lowland forests of Boulder River, waterfalls just starting to boom; and I think of a seldom-visited grove of redcedar along the upper Baker River, and the remnant old-growth roost trees used by bald eagles wintering on the Skagit above Rockport and Marblemount.

In the higher country of North Cascades National Park, there is the Big Beaver Valley with its famous cedar, which I first saw more than twenty years ago after days backpacking in the high mountains. There are also forests I have heard about but not yet visited: the extensive old-growth fir, hemlock, and cedar of White Chuck Bench east of Darrington and the giants of Noisy Creek at the north end of Baker Lake, the ancient stands along West Cady Creek, and others. Each stand is unique in its way, shaped by its own disturbance history and response to the topography, soils, and microclimate of its particular site. Each in turn supplies a vital and necessary thread in the larger fabric of the forest ecosystem.

When I cast about for someone intimate with the full range of Mount Baker–Snoqualmie forests, Jan Henderson's was the name most often mentioned. He is the area ecologist for both Olympic and Mount Baker–Snoqualmie National Forests. I knew of his exhaustive study of forest plant associations at Olympic, and also that he was at work on a similar study of the Mount Baker–Snoqualmie. After a few missed connections, I finally caught up with Henderson and his crew in the wet and rugged high country north of Big Snow Mountain in the Alpine Lakes Wilderness. For a couple of days we clambered up and down steep, off-trail slopes, coring trees and staking plots; then we rested in a high camp at Marlene Lake. There, Henderson and botanist Robin Lesher, who has worked with him since 1984, reflected on their work.

It was 1979 when Henderson began the considerable task of inventorying and classifying the full range of ecological communities in both national forests. To accomplish this, he, Lesher, soil scientist David Peter, and their crew established a network of over five thousand plots, nearly a thousand of them permanent sites that are revisited at least once every ten years. In total, these plots (each one-tenth to one-quarter acre in size) sample virtually every forested square mile of both national forests. Within the plots, the researchers record and catalog all visible fungi and vegetation, from trees and canopy epiphytes to shrubs, herbs, grasses, mosses, and lichens. They measure and core tree trunks to determine their ages, and they later study the cores under a microscope to assess each stand's fire and climatic history. They collect lichens and analyze them to detect the presence of heavy metals and other airborne pollutants. They note the presence, or sign, of all

Fungi fill many roles. They act as decomposers, cause disease, and live inside foliage, producing chemicals distasteful to insects. Thousands combine with algae to form lichens. Still others, including coral fungus, fuse with tree roots in a partnership that benefits both the tree and the fungus.

wildlife including invertebrates. Resulting data, so painstakingly garnered, are entered into a computerized Geographic Information System. What emerges after years of this work is a thorough characterization and description of the forests.

In talking to Henderson and Lesher about old-growth forests, I asked how Washington's westside forests compare with those of Oregon and British Columbia, and Henderson answered that there is a transition zone in the Mount Rainier–Snoqualmie Pass area. North of there, our wet, westside forests tend to resemble those of western British Columbia; cooler, moister conditions limit the intensity and extent of forest fires and tend to produce larger, older forests. South of there, the Cascades forests are more like those of Oregon.

"We've found lots of areas here that were undisturbed for 700 years prior to logging," Henderson commented. Did that contribute to Washington's high number of record-sized trees, I asked. Henderson shook his head. "We don't know what the real record trees are. They were cut down with the rest of the lowland forest early in the century, before anyone was keeping track." When I asked about mountain forests, Henderson and Lesher agreed that yellow-cedar are among the oldest and most impressive trees they encounter, but not necessarily because of size. Henderson found one eight-inch yellow-cedar that is 700 years old, a remarkable example of clinging to life with little girth to show for it. Diameters greater than three to four feet and ages of up to 1,000 to 1,500 years are not uncommon for this species, but age and size are not always positively correlated. On a poor site, aged trees may remain miniature.

But as much as any, it is the Pacific silver fir–western hemlock forests of middle and upper elevations that intrigue Henderson and Lesher. The silver firs themselves are not especially old, but the *forests* are. Having mostly escaped the large fires of the last seven centuries, many stands are a lingering legacy of response to changing climate. Scattered throughout the upper end of the silver fir zone are giant Douglas-fir 700 to 900 years old. These great-grandfathers are the final remnants of stands that seeded in after large burns ravaged forests during a climatic warming that lasted from about A.D. 800 to 1300 (the Medieval Optimum). At that time, according to Henderson, Washington's silver fir zone formed a narrow elevational band; Douglas-fir and western hemlock dominated the western Cascades, much as in Oregon today. Then climate started to cool as the Little Ice Age set in, and shade-tolerant silver fir seeded its way downslope, thriving beneath the older forests and in time replacing them. Their last downward step occurred about 200 years ago, as the Little Ice Age was coming to a close. Today, Henderson and Lesher commonly find silver fir 150 to 200 years old within the Douglas-fir–hemlock forest, but the trees are seldom doing well. They do not reproduce, and many are succumbing to insect infestation. Lingering at the lower end of their range, the trees mark the tailend of the Little Ice Age, a climatic shift five centuries long.

Mention of grand old Douglas-fir caught my attention, as it always does, and I asked where Henderson and Lesher encountered the most impressive stands. Beyond agreeing that the truly great stands are mostly gone, they answered that of those remaining, the Douglas-fir along Little Sandy Creek east of Mount Baker stand apart. Growing on ash-fall soils, these 500-year-old trees are five to ten feet in diameter and tower almost 300 feet high. Then Henderson mentioned another spectacular stand also on volcanic soil, in this case atop an old mudflow on the flanks of Glacier Peak.

The following evening I paused along the Milk Creek trail in Glacier Peak Wilderness to admire a recently fallen Douglas-fir. It measured nearly six feet through where it blocked the trail and had been cut. At that point (about twenty feet from its base) I counted 657 rings, though admittedly my eyes blurred over when I reached the close-set rings of the outer two inches. Regardless, my count indicated this tree sprouted following the large fires of about 1300 that Henderson had described. Other large trees—Douglas-fir and redcedar—were scattered throughout the forest, but not until the trail topped out on the first bench—the old mudflow Henderson had mentioned—did I wander off and lose myself in his grove of giants.

Large fire-scarred trunks of Douglas-fir shone faintly in the failing light, almost as if lit from within. Climbing over logs and stepping around pale blooms of rattlesnake plantain and bright red bunchberries, I was drawn from giant to giant amid the scolding chatter of Douglas squirrels. Eventually, at the toe of the mountain slope, I found the largest tree in the grove. With a diameter of nearly 8 feet, it rose almost 100 feet to its first limb; its crown lost somewhere above the forest canopy. I stood soaking in the presence of this grand old colossus and the squirrels quieted. Something of the tree's own stillness slipped into my very being. Toward dark, the sweet song of a winter wren followed me as I wound my way back to the trail.

The Forests of Eastern Washington

When traveling through forests east of the Cascade crest, we mossbacked westsiders are often surprised by the wonderful diversity of dry-side forests. There, familiar coastal-forest elements combine with continental forests from the Rocky Mountains and plants associated with sagebrush and the grasslands of the Columbia Plateau to create a complex mosaic. The ecology of eastside forests is born of wildfire and drought, and their character tends to shift suddenly with slight changes in elevation or direction of slope.

The twenty-mile drive from the grasslands of the Methow Valley to the montane forests of the meadow country around Tiffany Mountain makes clear the remarkable diversity east of the Cascades. Open stands of ponderosa pine and streamside aspen along valley bottoms give way to ponderosa mixed with Douglas-fir and grand fir as the road climbs Boulder Creek in the eastern edge of the Okanogan National Forest. Below Freezeout Pass the road cuts through fingers of lodgepole pine. Beyond the pass, an extensive lodgepole forest dominates the high plateau country. Engelmann spruce joins the pine in cool, moist draws, and wet sedge meadows ribbon the forest. The feeling is distinctly northern; this forest harbors the greatest concentration of Canada lynx in the state, a classic symbol of northern forest. Snowshoe hares, the lynx's primary prey, are keyed to lodgepole pine: they feed on the young tree growth that regenerates after a fire. But with a sixty-year history of fire suppression here, hares find less tender undergrowth beneath a closing canopy, and that trend is becoming a possible threat to the lynx population: less fire results in less browse for hares, which leads to fewer hares, therefore to fewer lynx.

Western larch, an eastside conifer, drops its needles following a spectacular October display of color. Fast-growing when young, the tree can quickly dominate a fire-cleared site. It does not tolerate shade well, so must stand taller than most of its cohorts, or die.

Such fire-related ecological linkages are vulnerable throughout eastside forests. For instance, it was fire that broke the dormancy of ceanothus seeds, a shrub that is a major contributor of nitrogen to the soil and a browse species for deer and elk. It was fire that maintained the open ponderosa stands preferred by flammulated owls, which are listed by the Forest Service as a sensitive species. And it was fire that created the snags needed by birds for feeding, perching, and cavity-nesting, and which after toppling serve as feeding and nesting places for creatures from wolverines to ants, as under-snow highways, and as enhancement for streams, creating the pools and riffles needed by fish and the invertebrates they depend on for food. In the past, wildfire played a central role in reinvigorating aging stands of lodgepole pines, and it also affected ponderosa pine and mixed forests. By eliminating the understory and thinning stands, fire lessened competition for moisture among survivors, which then put out new growth and were much less likely to suffer stress-induced insect outbreaks and disease than was the case before flames had swept through the forest. Frequent, low-intensity fires keep fuel from building up on the ground and thereby reduce the likelihood of major conflagrations, which are far more ravaging than "cool," low-intensity fires. Species like ponderosa pine, Douglas-fir, and western larch, when mature, have thick bark that insulates them from low-intensity fires; and many shrubs like ceanothus, mountain-ash, and red-osier dogwood, as well as herbs and grasses, resprout well after such fires.

The University of Washington's James Agee believes our recent human proclivity for

In northeastern Washington the Kettle River loops along the international border. The nineteenth-century trappers, miners, and loggers who used it as a travel corridor set the relationship between people and forest onto its present course, which now is nearly in crisis.

fire suppression, coupled with our selective logging of the most fire-resistant big, old trees has unwittingly traded frequent low-intensity burns for firestorms like those that blackened thousands of acres in eastern Washington during the summer of 1994. Human activity has also shifted forest composition from dominantly pine to Douglas-fir and true fir, trees that are relatively prone to outbreaks of defoliating insects, partly because of their overall genetic susceptibility and partly owing to their inability to recover from such attacks while also competing amid the moisture and nutrient shortages that follow a major fire. What to do about all this, however, is far from simple. Agee points out that returning to natural fire intervals means about every fifteen years for ponderosa pine, thirty years for mixed forests. That would involve control-burning more than 800 acres each year. As an alternative, some land managers propose a rash of timber-cutting prescriptions under the broad banner of forest health. Others, like forest ecologist Peter Morrison, fear such practices may be jumping the gun. A consulting ecologist with the Sierra Biodiversity Institute, Morrison has conducted ecological surveys for the Wenatchee and Okanogan National Forests and believes the "crisis management" approach advocated by some scientists and land managers is unfounded. And he fears that, over the long term, it may harm forests more than help them.

To see what he meant, I joined Morrison in the South Ridge area of Okanogan National Forest east of the town of Twisp. We walked into an open ponderosa pine–Douglas-fir forest on a dry south-facing slope; the forest had been partly logged. Looking at a 300-year-old Douglas-fir stump ten feet in diameter, Morrison pointed out that its rings showed no evidence of fire for close to 200 years. In another part of the forest, we found a ponderosa stump with evidence of a fire 22 years ago, preceded by a fire-free gap of nearly 100 years. Morrison's preliminary studies suggest that fire intervals of 80 to 100 years are not uncommon in these forests. He emphasizes that, even with our recent history of fire suppression, this places many of Washington's eastside forests well within the range of past fire intervals.

Unfortunately, on the east side, extensive research like this is years—maybe decades—behind research for westside forests. Morrison points out the tremendous need for baseline data and stresses the importance of determining the *longest* fire-free periods before we jump into what may be short-sighted approaches to forest management. Too many complexities are still overlooked, imperfectly understood, or even unrecognized. As one small example he pointed to the "witches broom" growth that had developed on a mistletoe-infected limb. "Forest managers tend to see that as a problem because mistletoe inhibits tree growth," he said. "But infected trees actually provide important habitat functions, especially when there's lots of mistletoe. Trees with mistletoe intercept more snow and provide better thermal cover for mule deer than those without it, and their limbs serve as shelter, nesting, and roosting areas for birds like warblers and spotted owls. We're just starting to understand the ecological contexts of these forests."

Earlier that same week I had explored old-growth ponderosa in the Pebble Creek drainage north of Twisp. Much as on Washington's west side, the east side's most productive forestlands occur outside national forests, and the great majority of them have been logged. Morrison estimates that 90 to 95 percent of our old-growth ponderosa stands are gone, but above Pebble Creek a few large cinnamon-barked "yellow-bellies" leaned among younger Douglas-fir, their old fire scars sealed over with pitch. As I sat among them on the warm, needle-littered ground I thought that given time, and help in the form of controlled burns, this resilient eastside ecosystem might heal its wounds. For if the history of Northwest forest management teaches us anything, it is to proceed slowly, retain options, and let age-old ecosystem legacies guide our actions.

British Columbia

By Yorke Edwards

Yorke Edwards is a biologist/forester who was raised in Toronto, Ontario, and there discovered birds when a sixth-grade teacher hung Audubon Society posters on the classroom wall. His professional career has been threefold: field research (on mammals) for the Parks Branch of the British Columbia Forest Service; interpretation, including development of the provincial parks' Nature House program and managing nationwide interpretation for the Canadian Wildlife Service; director of the renowned Royal British Columbia Museum.

Now retired, Yorke and his wife live "at the best birding site in all of Victoria," the low point of land where the Strait of Juan de Fuca meets the Strait of Georgia. He writes for and helps edit a variety of nature journals and his books include *The Land Speaks* and *The Mountain Barrier.*

Right: *Moist snowflakes first cling to twigs, then to each other.*

Reliant on moisture

blowing in from the ocean,

Sitka spruce pioneers

the entire Northwest coast.

It has an unusual tolerance for

salty air and buffeting by wind.

British Columbia

Meeting Old Forests

Forests are the ultimate expression of vegetation, and old forests are by far the best of all. This is especially true in the Pacific Northwest—which in Canada is the southwest.

Bound for the University of British Columbia as a graduate student forty years ago, I entered this realm of ancient coniferous forests for the first time. The hardwood forests of Ontario, for all their childhood familiarity and autumnal glory, had not prepared me for standing beneath the big trees of the western forests. It was an experience beyond words, though perhaps the word "stunned" comes closest. A similar experience might be that of a child from a prairie farm looking up at a skyscraper for the first time. But the forest is different. Its great pillars are alive, both near you and all around as far as eyes can see.

You walk *in* a forest, not *on* it as you do on a prairie or desert or most other landscapes. In an old western forest you are enclosed. You are a dwarf among giants, an ephemeral featherweight among tons of venerable living matter. Overhead are great volumes of living space with layer on layer of habitats for plants and animals. Looking up you seem to be standing at the bottom, yet underfoot are more volumes of living space, more layers of habitat. You stand on the top of composting levels, nature's recycling center. Unseen beneath you are myriad life-forms, most of them turning forest debris back into component materials, some of them preying on the composters, others simply taking temporary refuge within the rich mix. Everywhere are fungus strands and tree roots feeding on the compost. Ecologist Marston Bates points out in his book *The Forest and The Sea* that both realms manufacture food in their tops where there is abundant light and recycle debris in the dim light of the floor below.

When plants first produced wood, which gives great strength to stems and branches, trees became possible. Through countless ages, competition for light may then have led to taller and taller trees. Individual survival depends on not being too crowded by other trees. In the shade of neighbors, some trees die, others live but grow slowly in light too weak for optimal photosynthesis. Tall trees become winners; sunlight is strongest at the top of the forest, and by using its energy trees are the most successful earthbound invaders of the atmosphere. They create their own climate, quite different from that outside the forest. They cool warm air, tame winds, and pump water from the soil. They hold winter snowfalls aloft to be melted, delay the spring melting of accumulated snow on the forest floor, and turn fogs into local rain.

In Ontario I had experienced forests that were remnants, woodlots variously grazed and thinned for firewood, always surrounded by agriculture. A few forest patches within parks were old and intact, so in these I glimpsed the grandeur of southern hardwood forests with tulip trees and sycamores, the more northerly forests of beech and basswood, and northward farther yet, beyond agriculture, the long rocky hills that grow old yellow birch and sugar maple. But these failed to prepare me for the wonder of the forest legacy along the cordillera of the Far West.

My introduction was fleeting and mysterious, a few glimpses through a train window on my first entry into British Columbia. It was night, but as the train moved carefully west, down from Alberta and Kicking Horse Pass to the wet, west side of the Rocky Mountains, light from the window lit what seemed to be a continuous wall of forest. Then an unforgettable scene flashed by. The wall had changed. A new kind of tree was out there, the graceful, arching sheets of its foliage like a deliberate trim for the forest's edge. At the same time yellow flowers, large and almond-shaped, glowed below the trees. It was months before I knew that those arching fronds belonged to western redcedar and that the flowers were skunk cabbage. I had entered what is called the Western Hemlock forest, one of the twelve forest regions officially recognized in British Columbia's great expanse of 367,573 square miles, over twice the area of California.

On that train ride through most of the night, over three successive mountain ranges, into the valleys between them, and across the eastern part of the interior plateau, I crossed two-thirds of the province and without knowing it passed through five kinds of forest—some of them three or four times. Two forest zones were cold, high-elevation spruce forests, one was a wet and low-elevation forest of redcedar and hemlock, and two

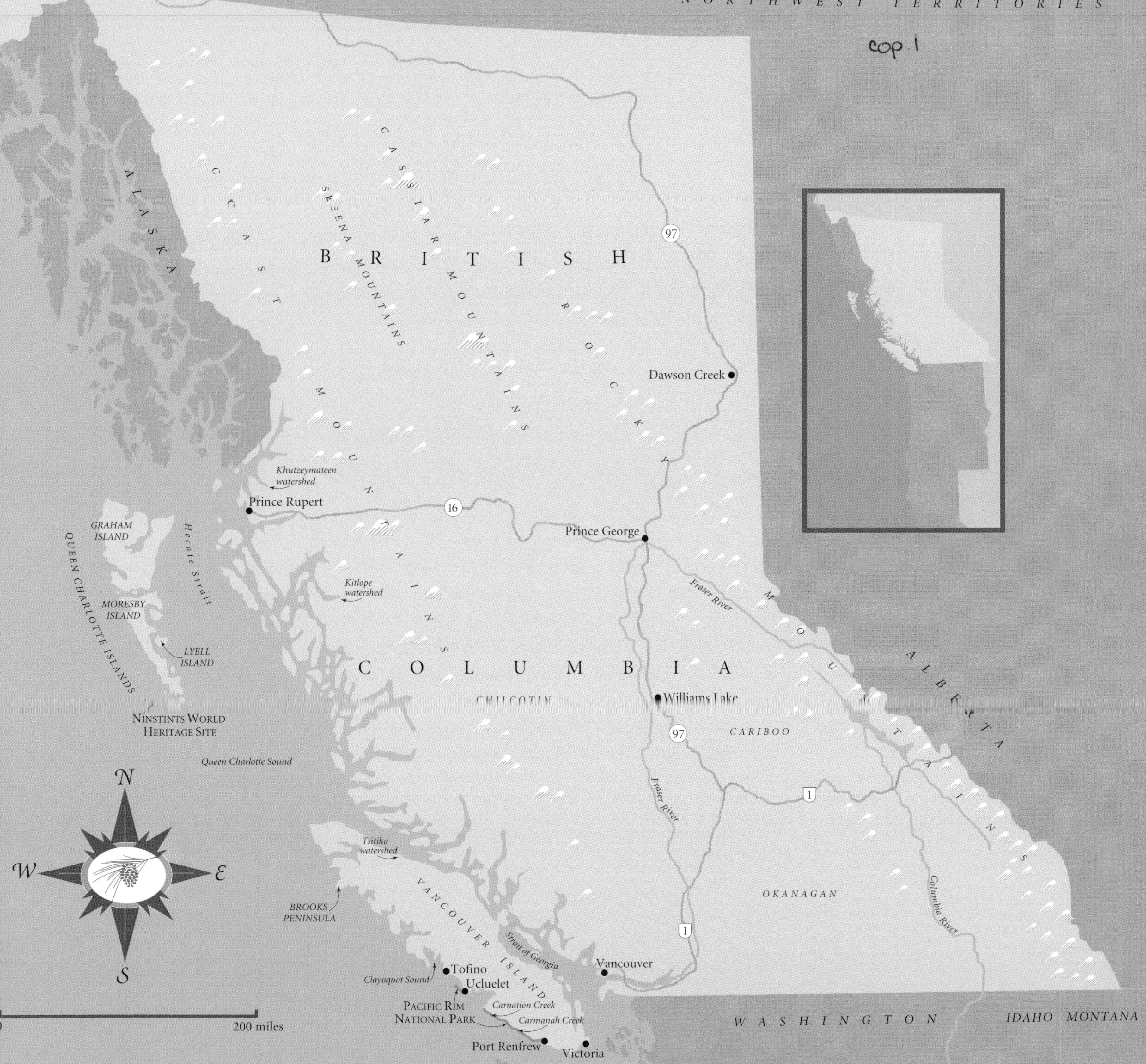

YUKON TERRITORY
NORTHWEST TERRITORIES
cop. 1
ALASKA
COAST MOUNTAINS
SKEENA MOUNTAINS
CASSIAR MOUNTAINS
ROCKY MOUNTAINS
BRITISH
COLUMBIA
97
Dawson Creek
Khutzeymateen watershed
Prince Rupert
16
Prince George
GRAHAM ISLAND
Hecate Strait
QUEEN CHARLOTTE ISLANDS
MORESBY ISLAND
Kitlope watershed
Fraser River
LYELL ISLAND
ALBERTA
CHILCOTIN
Williams Lake
NINSTINTS WORLD HERITAGE SITE
CARIBOO
Queen Charlotte Sound
N
W
E
S
1
Tsitika watershed
VANCOUVER ISLAND
BROOKS PENINSULA
OKANAGAN
Columbia River
Strait of Georgia
Vancouver
Tofino
Clayoquot Sound
Ucluelet
PACIFIC RIM NATIONAL PARK
Carnation Creek
Carmanah Creek
0
200 miles
WASHINGTON
IDAHO
MONTANA
Port Renfrew
Victoria

were dry, low-elevation forests, one ponderosa pine, the other Douglas-fir. At dawn I awoke in a hotel in tree-less grassland—cowboy country.

That first summer, working in the dry interior, the magnificent biodiversity of the province became clear in an unforgettable instant. Standing in sagebrush, I had just danced away from a defensive rattlesnake when my research partner pointed out a distant mountain still white with snow and told me that caribou would be there seeking summer snow patches. It was as if we were standing in Nevada but seeing into the Yukon. From tree-less desert we looked up into tree-less tundra. Between the two were bunchgrass flats, an open, grassy forest of ponderosa pine, and a wide belt of grassy Douglas-fir forest, which graded up into cold spruce forest that continued to timberline.

Skunk cabbage brightens ephemeral ponds and muddy places, a herald of spring in the soggy coastal forest of British Columbia. The province is so vast, however, that the forest of its coast and southwestern corner is only part of its environmental tale. Northern taiga and interior sagebrush are equally characteristic.

The wonder of that moment has not dimmed. I recall it often when I look south from Victoria across the Strait of Juan de Fuca to the snowy Olympic Mountains. I live in a patch of rain shadow that was grassland with oaks before it became a city. Yet across the strait I can see rain forests and subalpine forests, timberlines and tundras. It is like standing in Oregon and looking into Alaska.

The people of British Columbia impressed me, too, that first summer because of their obvious love of homeplace—a passion that has good reason. This land offers endless variations on the themes of habitat, abundant wild lives, weather, and vast expanse, all of which animate human lives. At the time, most of the province's ecosystems had yet to be severely modified or—in places—essentially destroyed. It was a totally fresh context for a young immigrant from the people-dominated flatlands north of Lake Ontario.

Perhaps in response to the many landscapes and seascapes that beckon and promise new experiences, British Columbians must be the most outdoor-oriented of all Canadians. People here write books about trees and eagles, deer and salmon, about escapes to Eden and how to become part of it. Elsewhere various degrees of urbanization seem to drive writers to fiction, perhaps because there is less to say about their real lives.

The legendary Cariboo and Chilcotin regions where I worked that first summer are rich in social history and widely read accounts of their character—and human characters. Both places have a large cast of footloose cowboys, remote homesteaders, Carrier Indian people, ranchers, packers, and summertime urban visitors who leave behind countless amusing stories to fill the long winter nights. They are the kind of places where biologists paddling back to camp from a hot day on alkali flats may find several pounds of moose meat hanging from the tent pole and never know who left it. Forest zones have distinctive plants and animals, and among the latter can be distinctive people.

The Forest Crazy Quilt

Why are there twelve kinds of forests growing in the province? The short answer has three parts: there are a lot of climates in British Columbia; each favors a different assortment of plants; and botanists have decided where it is reasonable to draw lines, although of course the vegetation assortments actually grade into one another rather than abut each other. The longer story is one of heat and cold, and mountains towering up for a mile or two. Huge weather systems with storm clouds carry tons of water from the ocean, which fall to Earth, and dry air carries away more tons of water from the land.

British Columbia is dominated by mountains; it is Canada's mountain province. The Atlantic provinces have the Appalachian Mountains, but they are so old their teeth have eroded to their roots. The scene in the Far West is different. From the Arctic Ocean into Latin America, the wide bands of mountains near the Pacific Ocean are young, and many peaks therefore are craggy and high enough to be snow-capped all year. In North America, mountain influence reaches far eastward across the Great Plains and westward to the Pacific. It strongly affects the life of the region, people included. Forests enter the equation largely because of climate, which is the overall sum of weather, which in mountainous country is in large part a product of terrain. Because different forests occur in different climates—some wet, some dry, some cold, some warm—mountains substantially rule on where a particular forest can flourish and where it cannot.

As seen from space the surface of British Columbia's southwest quarter has two ribbons of mountains roughly oriented north and south in line with the coast. East of each ribbon is a lowland, one narrow, the other a wide plateau. The most westerly mountains form the Island Range, which creates Vancouver Island, the Queen Charlotte Islands, and some islands off the Alaska coast. South from the Strait of Juan de Fuca this ribbon continues on the mainland; it includes the Olympic Mountains and the Coast Ranges which extend into California. The lowland east of these mountains is substantially flooded by the sea, waters that include Puget Sound, the Strait of Georgia, and a succession of other sounds and straits northward into Alaska.

The Coast Mountains in British Columbia form the second craggy ribbon, a distinct formation not to be confused with the Coast Ranges in the United States. Here, the Coast Mountains begin in the Yukon and extend south almost the full length of the province to the Fraser River. Just south of there the geologically different Cascade Range rises as a spine that extends into California.

East of the Coast Mountains is the dry country that first introduced me to British Columbia. It spreads as a wide and low plateau extending about half the length of the province and is divided geographically into three vaguely defined regions: the Okanagan in the south, the Cariboo to the north, and the Chilcotin to the west of the Cariboo. (Here, Okanagan and Cariboo are regional names never to be spelled "Okanogan," as in Washington, or "caribou," the name of North American reindeer.) These three areas are cattle country and tourist country, too, partly because of numerous lakes famed for Kamloops trout (elsewhere called rainbow trout).

Along with this geological pattern British Columbia forests are affected by three important weather considerations: ocean currents, elevation, and latitude. Ocean currents keep coastal lands unusually warm in winter for this latitude. They also warm the air, and as air warms, it becomes thirsty and sucks up vapor from the water. Ocean winds, warm and moisture laden, rise to cross the mountains. This chills them and they drop their water load, which drenches the land below with rain and blankets it with snow. Descending the east side of the mountains into the lowlands, the air again warms; becomes thirsty; picks up moisture; and holds it. Net result? The west sides of the mountains tend toward cloudy, wet weather especially along the coast, and trees there grow very large. The east sides of the same mountains have much cloudless weather and little rain. Some locations are too dry to grow trees, or grow only trees that can live on the limited water available in the rain shadows of mountains.

In high mountains, snow packed into ice forms glaciers, most of which are now retreating. Near the British Columbia–Alaska border, however, the Taku Glacier is advancing into the lowland forest and threatening to block salmon runs by damming the Taku River.

The second factor—elevation—is straightforward: high elevations are colder than low ones, and typically the forests of uplands are different from the better-known forests of lowlands. With shorter growing seasons, these higher forests grow more slowly; their trees are well adapted to "waiting out" the penetrating chill and heavy snow loads of winter.

As for the third factor, in our hemisphere north is colder than south. For British Columbia, which stretches north-south for 800 miles, this means that at about the halfway point cold spruce forests and other vegetation typical of the subarctic take over from classic Northwest forest. Within this spread of latitude, from balmy south to frozen subarctic, many tree species reach their northern limit. In many cases this seems to be a simple matter of increasing cold, possibly in others that the land has not yet been free of ice long enough for certain species to arrive. In still other cases, it may be that in recent centuries no great fire, infestation, or blowdown has sufficiently cleared-out competition and prepared the way for new species. Which form of disturbance might be expected to do this differs from coast to interior: blowdown for the coast, fire or infestation by insects for the interior; climate change for them both. For whatever reasons or

Coastal Douglas-fir is less tolerant of shade than its usual associates, western hemlock and redcedar. It therefore relies on blowdown, fire, disease, or other forms of disturbance to open the canopy and provide places where its seedlings can out-compete rivals.

combinations of reasons, nine coastal tree species that are common to the south do not extend beyond Vancouver Island: Douglas-fir, grand fir, white pine, Oregon white oak (called Garry oak in British Columbia), madrone (called arbutus, the genus name, in British Columbia), bigleaf maple, vine maple, western flowering dogwood, and cascara. Whitebark pine stops a short distance north of the island, and Pacific silver fir (in British Columbia called amabilis fir, the species name) gives up just beyond the Alaska border. On the plateau east of the Coast Mountains, ponderosa pine reaches northward for only a quarter of the province's length, and inland Douglas-fir reaches halfway, which is well north of all coastal Douglas-fir.

Three Companions

Three tree species have been my special companions in British Columbia. One is Douglas-fir, hard to get away from at low elevations over most of the southern half of the province. Another is oak, found only in small patches on the southern coast, where in places it is merely a mat two or three inches high, much as arctic willow may stand Lilliputian in size though hoary with age. My third tree companion is shore pine, not well known by this coastal name but inland called lodgepole pine, a common tree throughout the west from the Yukon to Colorado.

Douglas-fir, our most famous tree, thrives throughout the mountainous far west from central British Columbia into Mexico and from the Pacific Ocean east as far as Colorado. Close relatives live in California, Mexico, China, Taiwan, and Japan. The trees can be very old and, on the coast, very large. Reliable statistics are not nearly so dramatic as those in folklore and fuzzy memories, but even careful figures indicate that Douglas-fir often live 750 years and the oldest may reach 1,300 years of age, with heights reaching 315 feet and trunk diameters as great as 14 feet.

Researchers have found that Douglas-fir invaded the British Columbia coast about 7,000 years ago, after melting caused continental glacier ice to retreat northward. With some exceptional individual trees still flourishing at 1,300 years old, and if their ancestors lived equally long, today's oldest trees could be merely the sixth generation on our post–Ice Age coast. In this part of its range, Douglas-fir cannot live in the shade of other species and therefore often get their start after fires or other catastrophic events have cleared a growing site. In the comparatively dry rain-shadow country of the interior, however, the trees can compete successfully among other species and are considered a subspecies different from coastal Douglas-fir. The foliage of the inland trees is more blue-green, and their cones tend to be shorter, less than three inches long compared with up to four inches for the coast variety.

Most old Douglas-fir forests, both along the coast and in the interior, have been destroyed. They were easily logged. Open stands of trees and gently rolling terrain at low elevations offered ready access, and the hauling of logs to mills was mostly downhill, whether to the shipping lanes of the sea or the roads and rails of the Fraser River valley in the interior. The two forests—wet and dry—offered the best of woods, suitable for sawing into all reasonable dimensions including truly huge timbers. Both forests supplied the lumber that built the towns and cities of British Columbia, built the villages, barns, and grain elevators on the prairie, and through export became the construction wood of preference for the entire westernized world. Old Douglas-fir wood still shelters most British Columbians because most houses were made of it; Vancouver, for instance, is a Douglas-fir city. The wood's reputation as the best construction timber on earth derives partly, however, from the fact that world markets were supplied with the dense wood of ancient trees. Future Douglas-fir wood will be of lower quality and smaller dimension as tree farms cut juvenile trees only fifty to sixty years old.

The coastal Douglas-fir zone, its ancient forest now about 99 percent destroyed, once dominated a thin strip of land along the southeast coast of Vancouver Island and on many of the small islands off shore. That strip has a "Mediterranean" climate unique in Canada—mild, wet winters and dry summers. The interior Douglas-fir forest, perhaps "only" 70 percent destroyed, still cloaks much of the low-elevation plateau north about to Williams Lake, and the tree continues much farther north in other forest zones, although not as a dominant species.

My second tree companion, conspicuous where I now live, is Oregon white oak. These trees often grow mixed with Douglas-fir in the remaining forests of the coastal Douglas-fir zone, and there are also scattered patches of essentially undisturbed land with oak meadows, or just meadows without oaks. Victoria is at the driest end of a regional moisture gradient; it is subject to long summer droughts, which challenge even plants adapted to survive with only a few light summer rains. Official maps of the province's forest zones ignore the oak meadows except to nominally include them with Douglas-fir forest. This is a narrow view that disregards the affinity of oaklands in British Columbia with the oak/chaparral/grassland mosaic that is more or less continuous from Vancouver Island to California. These parched lands, lying in the rain shadows of the southern

Island Range, the Olympic Mountains, and the Coast Ranges, may have a built-in irrigation system if a recent study of oaklands in California is also applicable here.

The California researchers discovered something called "hydraulic lifting" in oak/chaparral. This is a subterranean, dark-of-night variation on the familiar daytime rise of moisture from roots to twigs to leaves and out into the atmosphere when sunlight is available for photosynthesis and leaf stomata are open and transpiring. Hydraulic lifting occurs at night and stops at the surface of the ground. It involves roots bringing water from deep in the soil and passing it to shallow, upper roots; from there it passes out into the soil rather than being stored until photosynthesis resumes. Both daytime and nighttime "pumping" are a matter of physics: moisture always moves from wet to dry. What is new is the discovery that substantial soil-moistening may result from hydraulic lifting. East coast research on sugar maple has found that "watering" by a single forty-foot tree may be as much as forty to fifty gallons per night. Furthermore, trillium and other plants growing beneath the tree clearly take advantage of this water, which is distinguishable from rainwater because of subtle chemical characteristics. Such revelations broaden our common idea of dry-habitat plants in competition for water. Sharing is also part of the equation. It may be that oaks in British Columbia, like those in California, foster their accompanying meadow community by watering it nightly.

No doubt people have deliberately modified oak meadows beginning long ago. Salish Indian villagers burned oak savannas periodically and dug their native root crops, in places most likely eliminating young oak and other trees; Camosun, the Salish name for the Victoria area, means "place to dig camas." Browsing black-tailed deer and Roosevelt elk probably also helped maintain the patchy openness of oaklands and meadows. Then came invading Europeans who farmed the meadows and used them to graze livestock. Indeed, the agricultural potential of southern Vancouver Island's naturally open land is what attracted the Hudson's Bay Company in 1843, the year it moved its regional headquarters north from Washington. The new Fort Victoria was the beginning of today's city, the driest coastal city in Canada and the one with the mildest winters.

Most remaining oak here far predate the lawns and flower beds they shade; some are dying because their root systems have been paved over or partly removed, some because today's homeowners find them in the way. Ironically, the oak/grassland community that attracted the first white settlement is now ranked as one of the province's most threatened ecosystems, and a Garry Oak Preservation Society has been formed. Lewis woodpeckers, once resident here, disappeared early in this century, and there is growing concern for Cooper's hawks and western bluebirds, species linked to open habitat. Among butterflies, a subspecies of the large marble butterfly has become extinct; a checkerspot has been lost to British Columbia; and the propertius dusky-wing is considered vulnerable because its larvae depend on oak.

Oak landscapes, like all the other components within our forest mosaic, have of course waxed and waned for thousands of years, keeping pace with climate changes since the retreat of glacial ice. Five to eight thousand years ago, when the climate was warm, oak was much more widely distributed than now. It was part of the hardwood forest that blanketed much of British Columbia in place of today's coniferous forest. As the climate cooled, oak continued to thrive on only the warmest, driest remaining sites, patches of land along the inner coast of Vancouver Island and its small offshore islands from Victoria north to about Comox.

My third tree companion comes in two forms: shore pine on the coast, lodgepole in the interior. The two are geographically separated but both are *Pinus contorta.* Proof that they are a single species is the usual one: they interbreed.

It was the early botanist David Douglas who bestowed the tree's Latin name, which means "crooked pine" and is appropriate for the coastal form, a tree seemingly unable to grow in a straight line. In high contrast, however, the widespread *Pinus contorta* of the interior—lodgepole—is famed for its straight trunks and its use as teepee poles by Native Americans. Indeed, over most of its range the "contorted" pine is exceptionally straight. According to the rules of botany, the Latin name of a plant species is a sort of personal logo that cannot be used for any other plant. The actual meanings of the words are unimportant, although in many cases they are usefully descriptive or historically meaningful.

This pine seems to grow best in very dry places and very wet places, in very low places and very high places. The trees often grace exposed shores where they are endlessly twisted by ocean gales, each pine a different shape, each somehow finding earth among seemingly naked rocks. On the wet outer coast, peatlands often miles across commonly cover flatlands and gentle slopes, and many are sparingly dotted with shore pines and yellow-cedars. Shore pine must have poor soil, a spartan requirement that serves as survival insurance in places where heavy rainfall constantly washes nutrients out of forest soils, a process called "leaching." Dry and rocky hilltops also have the pines. In mountains, hikers find them among the last trees at timberline. This species is obviously labeled "made to survive."

Its one great weakness, however, is that it cannot endure competition from most

other trees. It therefore grows in open places, often where the ground is either too dry for other trees or too soggy. Herein hides a mystery. The anatomy of many plant species living in peatlands is actually typical of those living in dry places, which seems especially strange as you slosh about looking at them. Part of the explanation may be that some bog plants, including shore pine, have fungi attached to their smallest roots, and these mycorrhizae may render small amounts of usable water from the overall acidic water of the bog. But this is speculation. The old mystery has yet to be solved.

The straight trunks of *Pinus contorta* in the interior may reflect habitat conditions rather than genetics. Lodgepole excels at rapid invasion of open land after wildfire, a two-part strategy that begins with their ability to seed burned areas ahead of possible competitors. The trees hold seeds at the ready year after year in cones that stay closed until opened by heat. Once the seeds are released, the strategy completes itself with copious germination followed by rapidly growing little trees, often so close together they shade out other plants and even stunt themselves. One scientist found an extreme example of this: a seventy-year-old "doghair" stand of lodgepole in which there were 100,000 trees per acre, all the same age, only about four feet high, and with trunks less than an inch in diameter. In normal, less crowded situations certain lodgepole grow faster than others and having reached the light of the upper canopy will dominate the forest for years until they are themselves overtopped and shaded out.

Shore pine and western redcedar form a distinctive forest at Long Beach in Pacific Rim National Park. Widespread soon after the last ice age, the pine now grows in scattered areas on poor soils. Its more familiar inland form, called lodgepole, ranges widely: it grows from the Yukon Territory to Baja California.

The Crazy Quilt's Past

Researchers have found that the climate has often changed in the past, producing corresponding changes in forests. For thousands of years glaciers covered most of Canada, including virtually all of British Columbia. Moving ice a mile or more thick squashed and sculpted the land. As it melted northward, thousands of life-forms began their reinvasion. In parts of the province a few small and hardy species had survived throughout the glaciation, but most species now in British Columbia pioneered the nearly lifeless land after the ice retreated. They came either from the south, beyond the continental glacier, or from unglaciated areas to the north in what is now the Yukon and Alaska. Such invasions were not smooth and continuous, for there were readvances of the ice within its overall retreat, and retreats of vegetation in response to the readvances.

Richard Hebda, a paleobotanist at the Royal British Columbia Museum, is among those who have studied the record of these vegetation changes, which is preserved in sediments in the bottoms of ponds and small lakes. There, in the mud laid down layer on layer through thousands of years is the pollen of once-thriving nearby plants. Sometimes they are identifiable only to genus, sometimes to species. Pieces of organic material, such as wood, are also present and can often be carbon-dated to reveal their age and thus the age of their level within the deposits. It is as if the date were written in the sediment.

One of Hebda's studies toward the upper end of Vancouver Island documented the parade of forests at that one location since the retreat of the ice. About 14,000 years ago the forest consisted of shore pine and alder, but very few other tree species. The climate was cool and dry. That this northern site was ice-free and tree growing at that early date came as a surprise. About 2,500 years later the pine and alder gave way to Sitka spruce and mountain hemlock as the climate grew wet and cool. By 10,000 B.P. (Before Present) western hemlock replaced the mountain hemlock in a warming climate; and by 8,800 B.P., Douglas-fir and Sitka spruce began to dominate in a climate that was warmer and drier than it is now. Then climate again reversed. Douglas-fir declined; Sitka spruce held its own, joined by western hemlock. By about 3,000 years ago the forest had become much like that of a century ago, when logging began.

Other research in botany now counters the old assumption that the province lay totally shrouded by glaciers during the last ice age. Scientists always agreed that a few mountain tops stood above the ice, but such places seemed unlikely to support plants through thousands of years of surrounding deep freeze. Certain plants, however, may now be telling us otherwise, for botanists are finding some of the same rare plants on the tops of six Vancouver Island mountains and in Washington's Olympics. They include one full species, *Aster paucicapitatus,* and several distinct varieties, for example, a paintbrush *(Castillija parviflora* var. *olympica),* a lousewort *(Pedicularis bracteosa* var. *atrosanguinea),* and a spring beauty *(Claytonia lanceolata* var. *pacifica).* Furthermore, there are at least three endemic species on Vancouver Island's Brooks Peninsula and in the Queen Charlotte Islands that are alpine plants but are growing near sea level: Queen Charlotte avens *(Geum schofieldii),* Queen Charlotte isopyrum *(Isopyrum savilei),* and Taylor's saxifrage *(Saxifraga taylori).* Apparently there were low-elevation places that were free of ice as well as certain rock peaks that jutted above it. All such locations are called "refugia." As usual in such instances, the picture is not clear-cut and future work will change our understanding. Most of the British Columbia coast has yet to see a botanist.

Actually, the first to suggest the likelihood of island refugia here was not a botanist but a zoologist, Bristol Foster. Studying mammals on the Queen Charlotte Islands, he found a number of species which are different from those on the mainland. For example, black bears in the islands proved the biggest in North America, a status also true of

An incomplete, long-abandoned canoe in the Queen Charlotte Islands could well be made from one of the original cedars to reach the archipelago. Men left such canoes over the winter to "season" and reduce the likelihood of splitting when the canoe was skidded to the beach for final carving.

these islands' martens and river otters. In addition a weasel, a shrew, and a deer mouse differ enough from those elsewhere to suggest that isolation on the islands has caused evolutionary changes. Recent studies show that these changes took place after the ice retreated. As Darwin first noted, evolution creeps ever onward, altering the legacy successive generations inherit. As usual, this truth is most evident on offshore islands.

Within the sweep of time, what happens at one place may differ substantially from what occurs at another place. The pollen data Richard Hebda used to determine the reinvasion rate of western redcedar illustrate the point. He finds no major evidence of the tree in northern Washington or southern British Columbia 7,000 years ago or earlier, perhaps because of unfavorable climate. But by 6,000 B.P. redcedar had become abundant in northern Washington, and pollen samples from the Fraser Delta indicate abundance there by about 5,000 B.P. However, at Prince Rupert (near the Alaska border) redcedar was not abundant until 2,500 years ago. It had taken 4,000 years to migrate about 600 miles. Seeds can travel but only via unreliable means of transportation: whether wind, or water, or animals. This particular journey involved crossing wide rivers and fiords. Hard-to-reach islands evidently—and understandably—posed an even greater challenge. Redcedar did not reach the Queen Charlotte Islands until 1,000 years ago. This means that some of today's big cedars in the Charlottes may be first-generation immigrants. Many must be second-generation.

At Ninstints Haida mortuary poles memorialize the high-ranking dead whose mortal remains they once held and also bear witness to the dislocation of a rich culture.

Euro-American culture on the Northwest coast has been heavily dependent on Douglas-fir trees, but before these newcomers arrived, it was redcedar that supported human cultures in much of the Pacific Northwest. At the tidewater edge of the great forests, the people of the First Nations developed highly successful cultures, drawing from an environment with wet and stormy weather, angry seas, whitewater rivers, and forests that grew more living tonnage and dropped more debris than any other forest in the world. Evidence of their villages can still be found: perhaps a tilting cedar house post with a cedar plank still attached to it or a decaying cedar totem pole fallen into the salal, its carved faces worn and cracked and only faintly visible in slanting light; perhaps a cluster of great depressions in the forest floor, each the imprint of a house.

Remnants of one particular village in British Columbia, abandoned a hundred years ago, have been designated by UNESCO as a World Heritage Site. This is Ninstints—or more properly, in Haida, *Sqa'ngwa-i Inaga'i,* which means "Red-Cod-Island-town." The village crowds the shore of a quiet bay on Anthony Island (Red-Cod Island), facing away from the sea's storms at the south end of the Queen Charlotte archipelago. Studies by staff from the National Museum of Canada show that this village once had sixteen houses and about forty-three totem poles, half of them mortuaries that held burial boxes. More than two dozen poles remain, many still erect. Huge fallen roof beams and a few upright posts and planks augment depressions as indication of houses.

To stand amid such remnants and realize what has vanished is to feel shivers the length of your spine. The people here—and along the entire Northwest coast—were magnificently successful. They built oceangoing canoes by hollowing cedar logs. They preserved food in summer for use in winter. They created awesome cedar architecture, a rich array of memorized story and myth, theater arts and costuming that rank high on a world scale, and artistic traditions strong enough to survive Christian and government persecution and enter a renaissance now internationally celebrated.

Salmon: Product of the Forest

All along the Northwest coast, salmon swarm into rivers and creeks to spawn, in places swimming hundreds of miles inland from the sea. They were an essential resource for aboriginal people, and their runs were major annual events—with an occasional poor run foretelling a hungry winter. These red-fleshed fish grow large in the nutrient-rich waters of the North Pacific, but they spawn in freshwater. There, new generations hatch and, from their first bite to their last one before entering the sea, they feed on forest nutrients cycled through a host of aquatic insects in all stages of development. Salmon are a product of the forest.

But this pattern, thousands and thousands of years old, may now be jeopardized. In the century and a half that Euro-Americans have dominated the Pacific Northwest, changes to the land and its river systems have taken place on a scale inconceivable to the people of the First Nations. Silt and debris, largely from logging, have sterilized creeks and rivers by eliminating species dependent on high-quality aquatic environments, and dams have destroyed populations of migratory fish by blocking the way to traditional spawning beds. Entire salmon runs have perished; others are now much depleted. A few enormous runs still flourish, however, and mask the overall trend. Today trouble in the salmon fishery is serious enough that all taking of certain species in certain waters has been forbidden in the hope that recovery will result. The fish's forest habitat has been gravely harmed, and too many fishing boats equipped with astonishingly sophisticated gear have taken too many salmon at sea. Nations around the North Pacific Rim now battle with words and on the water over sharing the remaining stocks.

On the west coast of Vancouver Island, scientists from Canada's Department of Fisheries and Oceans have studied some aspects of the Carnation Creek watershed for nearly twenty years. As is typical of the region, this creek drops rapidly to the sea from its rise just a few miles inland. Research at the creek started well before the watershed was logged and continued during logging and for several years afterward. Parts of the forest were not logged. In other parts, trees were cut right to the water's edge. In still others, forested strips up to 230 feet wide were left intact along the creek. Some logged areas were burned to remove debris, an established practice, and others were treated with herbicide to remove unwanted vegetation that might compete with plantations of seedling trees, also an established practice. Researchers recorded the stream effects they observed—and no doubt missed others. Ecology has been characterized as not only more complicated than we think but quite surely more complicated than we can think. Even so, a summary conclusion stands out clearly in this instance: logging was a disaster for Carnation Creek's salmon population.

Two-inch chum smolts signifiy the essential oneness of forest, creeks, estuaries, and ocean. In the words of Vancouver Island Nuu-chah-nulth elder Simon Lucas: "If the water can no longer support the salmon, and if the land can't support the deer and bear, then why do we think it will support us?"

Habitat changes combined to cause the loss. With most of the forest gone, the soil's capacity to store water diminished so markedly that on occasion storm runoff nearly matched storm rainfall in volume. Severe channel shifting and erosion resulted, and streambed deposition of silt, clay, sand, and pea gravel increased greatly. For instance, the proportion of sand from cutover areas doubled, degrading the quality of salmon spawning beds, or even destroying them. The number of logs in the creek, necessary to fish as cover and for stabilization of banks, declined 30 percent and remaining logs tended to concentrate in large, unstable piles. Long straight "glides" of water increased, riffles decreased.

Stream nutrient levels were also greatly affected. They declined drastically two to four years after logging and herbicide use, although at first they had increased, at least during periods of high flow. Initially a greater amount of fine forest debris, part of the invertebrate food chain, was present in the water, but it was flushed through and lost. Leaf and needle litter similarly dropped to half, or even to a quarter, of what it had been before logging. The impact of this is substantial: stream nutrients nourish aquatic plants, which feed insects and other invertebrates, which feed fish. Yet after logging, the larvae of mayflies, midges, gnats, and stoneflies—the principal food species for young salmon—declined 40 to 50 percent.

In parts of the watershed that had been logged, stream temperatures increased by as much as five degrees Celsius (nine degrees Fahrenheit), a marked rise that was sustained throughout the postlogging years of the study. After logging, the intensity of the light reaching both stream and slopes at least doubled; shade was all but gone.

How all this affected fish differed by species. Sculpin vanished where clearcutting had denuded stream banks, and they declined in the watershed as a whole. Cutthroat trout showed little change. Steelhead declined. Chum salmon, genetically coded to spawn a short distance upstream from the ocean, reacted dramatically. Production of eggs and young immediately fell, and the number of adults returning to spawn dropped 95 percent. (Some of this, however, could have come from unusual changes in ocean conditions concurrent with the study period.) For coho salmon, the species with the most data, logging at first seemed to be a benefit. The warmer winter and spring water temperatures fostered earlier hatching and therefore a longer period for growth. That meant larger fish by fall and better odds of their surviving winter, for coho remain in freshwater about fifteen months before going to sea. The increased size and enhanced winter survival lasted for only six years, however. Thereafter, each year the survival rate for first-year coho fell below the prelogging survival rate. Coho already in their second year of life when logging began immediately suffered more than a 50 percent loss, and after six years such fish became rare. Logging triggered decimation. Damaged forests produce damaged salmon runs.

Our Disappearing Old Growth

Forests, like wine, improve with age. In both, a good vintage can be a rich biological experience, memorable because enjoyed and also pleasantly mysterious because age has endowed both wine and forest with an excellence not yet fully understood.

Numerous species live in old forests. In comparison, a younger forest is biologically spartan, and once the canopy closes, the gloom beneath it reduces most other life. This huge difference between old and young forests is at the root of British Columbia's current confrontation between industrial corporations that log and sell wood, and citizens who want old forests preserved. Understanding the situation requires some knowledge of British Columbia's peculiar system of forest management.

Across Canada, confederation created a forest-management base different from that in the United States. The national government owns no forests within the provinces.

When the colonies joined to form Canada, all public lands and forests (and other resources) were given to the provinces. Since then, most provinces have sold little forest land. The result is that on paper they still seem to control about 95 percent of Canada's forests south of the two northern territories. The other 5 percent is federal land acquired by agreement with the provinces for national parks, military reserves, and other public needs.

Within the provinces, however, the reality is that trees have been virtually given to industrial corporations. As a result companies selling wood are among the biggest and most politically powerful nongovernment organizations in the nation. In effect, British Columbians do not own the trees on most public lands. Industry has some obligations, but many of these are easily modified and so relatively inconsequential. Eight companies own logging rights for about 50 percent of the province's forests, and about 70 percent of all logging profits leave the province. Operating costs, including wages, do stay in the province, of course, and this seems to be the main justification for what otherwise is government's curious "management by nonmanagement" of public wealth.

Along much of the coast, companies have been given tree farm licenses, which are good for twenty-five years and renewable repeatedly, subject to approval. Most tree farm licenses cover large areas, up to nearly two million acres, and some companies own several licenses. These are worth millions of dollars and can be sold to the highest bidders. The logging industries essentially control the forests and their fates, and the provincial Forest Service usually agrees to their requests—even to bending the rules. The government does receive annual revenue from the companies on logs removed from the forest, a sort of "stumpage tax," but its rate fluctuates widely depending on how the market is currently affecting the companies. The Forest Service is the only government department given jurisdiction over public land. Other government offices, like fisheries, wildlife, and recreation are poor cousins by comparison, with little real management power.

Using its ability to direct major forest decisions, industry in recent years has increasingly enlarged the annual cut of trees in the province. The goal of increased revenue has been allowed to override what textbooks detail as the very soul of good forestry, namely that the annual volume of wood cut must not exceed the annual volume grown. This policy ensures a sustainable cut in perpetuity, hence sustains forest-dependent jobs and communities in perpetuity. Overcutting, in contrast, is simply mining the forest—ultimately a guarantee of job losses and communities in decay, not to mention harm to the planet.

This industrial freedom is the cause of growing discontent in British Columbia. Shortsighted, money-oriented thinking has prevailed, except in parks, which history teaches are not completely safe from logging. Environmentally oriented organizations, especially on the coast, are countering this trend, capturing increasing media attention with their protests and causing rising concern in government and industry. A response has been the creation of dozens of new provincial parks, most in small patches, many with surviving remnants of our forest legacy. This "solution" tries to benefit both sides of the issue equally. Predictably, only moderates are satisfied—and neither most of the logging fraternity nor many environmentalists are moderate.

Citizens' physical confrontations have been largely confined to coastal forests except for periodic short demonstrations at government buildings. Outrage is based on two areas of discontent: the preserving of too little rain forest and the use of clearcutting as the universal logging method. Both trends spell continued decline for the province's heritage and biodiversity, for there is no rare-species law in Canada to help protect wild ecosystems.

There is an industrial bonanza here. Beyond the costs of cutting and hauling, our trees come free for the taking. They are wooden gold, ancient timber of top quality and huge dimensions sold for top dollars. There is also a forestry problem. Trees planted after clearcutting, then cared for by silviculture, cost money. If cut young, they yield inferior wood of low market value, but depending on a host of variables from tree species and soil types to market fluctuations, the expense of caring for them longer eats up maximum return on investment.

The basic fact is that after about sixty years the value of well-invested money is likely to grow faster than that of additional wood, although obviously no one can really know the future. Even so, at a certain point, the growth rate of invested dollars is likely to leave that of trees behind and that point comes in perhaps sixty years. To illustrate: in sixty years a dollar invested at 10 percent will be worth $304 and if left another twenty years it will compound into $2,048. With trees, however, after about sixty years the added value from further growth is unlikely to match the rate of gain attainable by the invested dollars. This basic reality leads companies with tree farms to rely on accountants, not foresters, to decide when to cut trees. Sheer economics tend to be against continuing to tie up capital by growing trees for the additional twenty to thirty years it takes them to mature, let alone for another century and a half until they reach the stage defined as old growth. Tree farmers using clearcut methods cannot afford to produce high-quality, old wood if profits are the central purpose. Inferior wood from young trees, probably held together by glue, may be the dominant wood of the future.

Vast peatlands with stunted forests form where poor drainage creates ponds and favors a succession of sphagnum mosses. Such forests often figure prominently in land preservation designations, but actually they are worthless for timber production, in any case. Peatland trees survive on hummocks [illegible]

The Peaceful Fight for Clayoquot

In 1988 citizen protests against logging practices began to focus on Clayoquot (pronounced "Klak-wit") Sound on the west coast of Vancouver Island, one of the last large areas of ancient rain forest in the province. This focus culminated dramatically in 1993, when for months the name "Clayoquot" was international news. Protesters built a large summer camp beside the only logging road into the forest, and almost daily small portions of the crowd blocked loggers from their work. A court injunction banned such blockades and declared any future ones in contempt of court.

Nonetheless, protests continued. Through the summer over 850 people were arrested, the largest-ever mass arrest in Canada. Trials clogged the courts all through the winter and spring—so far as is known, the largest mass trial in the world to date. Scores of those arrested served jail sentences rather than agree to remain away from the blockade. Hundreds occupied the camp all summer; overall, because of constant arrivals and departures, thousands were involved. Protesters were mainly young adults, but children were arrested, too, and soon released; grannies were jailed, some insisting on it when offered the alternative of confinement at home. People came to the blockade from across Canada and from other countries. Clayoquot was an international issue.

That summer of civil disobedience followed years of media experience gained through other protests in other forests. The approach, organized primarily by the Western Canada Wilderness Committee of Vancouver (WC^2 for short), was to show what we are losing, reveal how little we know about it, and urge people to inform their government of their concerns. A virtual ton of letters to politicians resulted. Among the strategies used to engender the letters were building good trails into the forest so that people could experience the specific area and publishing well-designed tabloid newspapers portraying the Clayoquot forest, as well as a superb coffee-table book celebrating the beauty of the forest and its island-studded waters. A program to create new knowledge about the area was also begun, and continues. It attracts scientists to both the forest itself and to research programs available through the Clayoquot Biosphere Project, a nonprofit, community-based environmental research organization. So far studies have focused on microclimates, forest-canopy lichens and invertebrates (mostly insects), marbled murrelets, winter birds, black bears, aquatic invertebrates, and salmon fry habitat.

Early in the Clayoquot controversy, the provincial government created a study group composed of local people and carefully chosen experts to work out a land-use plan. Washington, D.C.-based Conservation International helped, as did other organizations

Steep-walled Sydney Inlet is representative of Clayoquot Sound's serene beauty. An early 1990s controversy over logging there led the provincial premier to impanel scientists and Nuu-chah-nulth leaders, charging them to come up with a plan that would "result in the best [forest] practices in the world."

and individuals. The planning made progress, although not on the main issue, which was "where to log, where not to log." Solid stalemate ensued. News media on the coast reported both sides of the whole issue, usually showing by the extent of their coverage that forest savers held the most mass appeal.

In 1993 a tentative land-use plan for Clayoquot watersheds, based heavily on public input, was released by the provincial government. It pleased neither loggers nor savers, and everyone agreed it tried to give half of "the baby" to each side. Forty-five percent of the land would be industrial forest; 33 percent, preserved; and 18 percent, "special management lands," to be logged with concern for wildlife, recreation, and scenic beauty as viewed from the water. As maps make clear, this decision fragments much of the remaining rain forest into a series of patches, so much so that many plants and animals will be at risk. A possibility, however, is to connect the patches, a solution that may work if logging on the special lands avoids clearcutting. It should be noted, too, that from a forest preservation standpoint the plan's land percentages are somewhat misleading. Much of Vancouver Island is tree-less alpine and subalpine country, exposed rock, and stunted forest on poor sites. The industrial forests' 45 percent embraces few of these lands.

As a next step in solving the Clayoquot issue, the provincial government selected an independent panel of distinguished scientists and Nuu-chah-nulth leaders who were asked to study the lands that had been designated for industrial forestry and to recommend practices that would assure sustained production. The panel's final report was issued in April 1995 and was accepted by the government. The panel pointed out that although maximizing timber yield has been industry's one goal, forests actually embody other values as well. They provide ecological services such as clean air and clean water. They hold specific objects of high interest such as huge, old trees and unique species. And, especially for First Nation people, they harbor places of special spiritual and cultural significance.

The Rising Tide of Dissent

The Clayoquot protest was based on rising public discontent that began in the 1970s. Before then Vancouver Island's great legacy of ancient forest was accessible only by one or two notoriously bad roads. Then Pacific Rim National Park with its West Coast Trail from Port Renfrew to Bamfield opened the lower coast of the island to the world of wilderness hikers, both local and international. The truth was out. Furthermore, the trail's opening roughly coincided with completion of a paved road to the central coast and the villages of Tofino and Ucluelet. With the road came easy small-boat access, and that plus the trail revealed the wonders of the forest. The drive to the coast, however, led through miles of stumps and other wreckage in centuries-old forest stands.

Public reaction set off British Columbia's seemingly annual series of forest blockades. The first began on November 21, 1984, on Meares Island (near Tofino) when loggers arrived in a boat to begin cutting. They never got to the forest. A flotilla barred their way by water and a human throng blocked the shore. The Nuu-chah-nulth people were protecting their ancestral home.

With traditional hospitality they welcomed the strangers to the island, but only if they would leave their chain saws in the boat. Two days later, Canada's biggest wood products corporation sought a court injunction; the islanders then filed proceedings against the corporation. The matter was in the courts. In time, an appeals judge favored the islanders, but with a ruling that prevails only until overall land claims can be addressed. Settling that may take decades.

During the long wait for a second court appearance, the Western Canada Wilderness Committee issued and widely distributed a tabloid newspaper reviewing Nuu-chah-nulth history and plans, and they also published a thin but high-quality book clearly describing the island's assets and the islanders' case. Public reaction to all this caused the news media to discover that forest blockades are front-page material.

Saving South Moresby, the area on and adjacent to the southernmost of the Queen Charlottes' two main islands, boiled into a public issue in 1985; loggers were about to denude the slopes around Windy Bay, an especially scenic area on Lyell Island, offshore from Moresby. From the beginning it was mainly a national issue in need of provincial cooperation. The goal was a national park. In support, an antilogging protest started, as on Meares Island, by blocking cutting crews' access to the island.

Dozens of Haida elders sat calmly on chairs, wearing traditional dress. Dozens were arrested and refused to agree not to blockade again. The heat was on, and it was fanned by Western Canada Wilderness Committee tabloid newspapers, detailed coverage in national and international magazines, a train loaded with park supporters traversing the nation and gathering national support, illustrated lectures across the country (with special showings to decision makers in Ottawa), plus—quietly—a massive network of societies, governments, and individuals at work convincing leaders in both provincial and national governments to save the island. It worked. There is now a South Moresby National Park Reserve, which the Haida call *Gwaii Haanas,* "Place of Wonder." It is still not fully a national park because of problems slowly being resolved, but even so it protects 358,000 acres of forest and waterways.

Next came the Carmanah Creek watershed confrontations, which dominated the province's "save-the-rain-forest" scene from 1988 until an even stronger focus targeted Clayoquot Sound. In Carmanah, the Western Canada Wilderness Committee turned on all its peaceful guns. Artists were invited to camps among some of the province's biggest trees, and subsequently an art show traveled the province, its canvases for sale in silent auction. Additional revenue came from a large award-winning book with 138 reproductions of the paintings accompanied by artists' written statements. Good trails were built so that the public could see what was about to vanish—and industry's chain saws could be heard from the trails. Bus trips were organized. Mass mailings of tabloid newspapers went out. Research projects got underway at a new study center.

The idea of creating facilities to attract scientists into the Carmanah was brilliant, and it characterizes the Wilderness Committee's long history of successfully drawing money and good people to well-planned projects that outclass government and industry. For example, marbled murrelet studies have established the little sea bird's use of this forest for nesting; estimates of the population involved; and some idea of the nesting density—although the word "nest" conveys the wrong image. Marbled murrelets literally balance their one egg, and later their chick, on the lichen-padded limb of a giant tree, which makes finding their nest sites notoriously difficult. Another research example, from the forest floor, is the discovery that salamanders are not only numerous but, with their penchant for eating almost any little animals they can swallow, may collectively out-eat other carnivores present in the rain forest.

Insects hitherto unseen—by humans—also are a part of the Carmanah studies. Neville Winchester, of the University of Victoria, has collected 1.2 million specimens of arthropods (largely insects) taken at ground level and from removable platforms that are 130 to 200 feet above the ground, strapped to catwalks that link five Sitka spruce trees. Of these one million-plus specimens, 150,000 were sent around the world for identification, and the first 40,000 to come back included 1,300 named species with 67 of them new to science. Winchester expects that by completion of the classification process, 600 new species will have been discovered. Already the research has given new understanding of insects' extremely complex predator-prey relationship, one of the ancient forest's major ways of warding off infestations. Furthermore, Winchester's research provides the first solid evidence that several arthropod species are so strongly linked to old growth that they will probably become extinct if all of the old-growth forest is cut.

None of this research has been universally welcomed. In 1990 vandals burned the newly built research camp and cut up miles of public boardwalk into the forest. Thereupon the Western Canada Wilderness Committee demonstrated its astonishing vitality. Volunteers simply rallied and rebuilt both camp and trail. Eventually the government proposed that half of the watershed be preserved, the rest logged. Then in 1994 it declared the entire watershed preserved—success again at saving a remnant.

Immediately south of Carmanah is the Walbran watershed, another intact patch of rain forest. Recently it, too, was declared entirely preserved, but only after public outcry. Following the initial, partial success of Carmanah's preservation, an intensified focus on the Walbran forest began in 1990 with road blockades, court injunctions, civil disobedience, and arrests. This seems to have been the first confrontation to attract people from beyond Vancouver Island to actively participate in a forest protest. Recognition that Canadian rain forests are global treasure had begun.

It continues. Names like Tsitika, Khutzeymateen, and Kitlope now fit within mainstream awareness. In the Tsitika watershed the forests that engendered protests against logging were the only extensive coastal Douglas-fir old growth remaining on the drier and more accessible eastern side of Vancouver Island. Protest focused on both trees and orca whales, with emphasis quite properly on the whales. Robson Bight, the shores and waters at the estuary of the Tsitika River, is famed for attracting orcas, which rub against the shore, apparently primarily to rid themselves of barnacles. When a logging company

Ecological linkages between death and life, plants and fungi, animals and microorganisms characterize old-growth forests. They are fundamental relations, yet only beginning to be understood. The linkages are both ancient and intricate, and scientific research concerning them requires long-term social, political, and financial commitment.

prepared to enter the area, concern over disturbing the whales and ending their use of the shore stirred public objection. Again there were blockades, injunctions, arrests (forty of them), a colorful tabloid newspaper, a high-quality book, and artists invited to painting sprees. This time the logging company countered with an expensively produced televised movie. The final outcome is a wide buffer of forest preserved along the shore and inland patches of undisturbed old-growth forest, in all about 3,000 acres.

Preserving the Khutzeymateen watershed was not so much an issue concerning old forest as it was the preservation of an unusual population of grizzly bears. A publicity campaign began in the early 1980s and led to the establishment of a provincial park about ten years later. Located north of Prince Rupert and across a fiord from the Alaskan border, Khutzeymateen was too remote for on-site demonstrations. Nonetheless, research information and views from chartered boats fed newspapers a steady stream of feature stories and kept the issue alive. Rumor has it that public expression of concern boosted the final preservation decision, which officially included acknowledgment that logging the Khutzeymateen was economically unsound, in any case.

The Kitlope watershed, part of a huge area of untouched rain forest (1,235,500 acres), more than half of it the Kitlope Valley itself, was declared preserved in 1994. Beginning about 1990 this drainage at the head of Gardner Canal, a long fiord on the northern mainland coast near Kitimat, received enthusiastic attention from many of the people most informed and discriminating concerning temperate rain forests. No confrontations occurred. Haisla First-Nation people simply voiced strong objections when the forest and waters they claim as theirs seemed about to be invaded by loggers, and they spurned the willingness of the company holding the cutting license to capitalize a separate division with Haisla people doing the logging and hauling. At about the same time Ecotrust, an organization based in Portland, Oregon, and dedicated to ecosystem preservation, recognized the Kitlope as one of the largest and most intact temperate rain forests in the world. As pressure mounted, both government and industry decided to show themselves responsive to public opinion.

The Haisla people never relinquished their aboriginal title to Kitlope. Therefore they insisted it not be logged and consulted Ecotrust and forest scientists. The resulting plan calls for managing a nearly one-million-acre wilderness for Native cultural and subsistence use, scientific research, and ecotourism.

As a result, think tanks are now pondering what will permit the Haisla people to both use and preserve their forest in this modern era. The vision is to keep the forest intact, including its element of human resource extraction as traditionally practiced. Kitlope may demonstrate how to perpetuate a forest ecosystem with aboriginal people still a part of it despite their having one foot planted in industrialized society. Planning moves toward the concept of successful preservation with gentle use.

The tale is not just one of politics and economics. It is also the story of a race with time. Kitlope may point the way.

Southeast Alaska

By Richard Carstensen

Richard Carstensen first came to Southeast Alaska from the East in 1977 "to experience truly wild land," and that land is now his stock, trade—and passion. He interprets it through writing, pen-and-ink illustrations, teaching, environmental consulting, map-making, and guiding. Since 1988 he has worked for Discovery Foundation, a nonprofit organization providing natural history education to the youth and teachers of Southeast Alaska.

Richard lives near Juneau on the shore of Auk Bay. From his windows, the Admiralty Island National Monument/Kootznoowoo Wilderness is visible in the far distance on clear days. He is the illustrator and a co-author, with Rita O'Clair and Robert Armstrong, of *The Nature of Southeast Alaska: A Guide to Plants, Animals, and Habitats*, a concise yet comprehensive treatment of the area's flora and fauna.

Right: ***Tongass National Forest***

Marine resources in Southeast Alaska are so rich that shells pave many beaches, and creatures that belong to the forest belong also to the shore. Black bears squish open barnacles and eat them. Brown bears occasionally dig clams. Shrews and mink feast on sand fleas, mussels, and snails. Deer graze the sedges of tide flats.

Southeast Alaska

Crowded Remnants: Prince of Wales Island

I wake beneath the spreading branches of a beachside Sitka spruce. My camp is midway between a goose nest and a bear's kitchen; I hope far enough from each that neither creature spent the night dreaming about me, as I did of them. At midnight the goose couple quietly discussed something in the sprouting sedges twenty yards below my tent door. The big black bear had been grazing at the creek mouth as I unloaded my kayak. Now, at first light, my food bag still hangs intact from its branch. Fortunately, the bear was content with wild foods.

This is Sea Otter Sound, an entrancing island complex off northwestern Prince of Wales Island, in southern Southeast Alaska. The island is the nation's third largest, after Hawaii and Kodiak. It is also the heart of Alaska's timber industry. Kayaking its western shorelines keeps me oscillating between euphoria and sadness, according to the success or failure of land managers at hiding their activities behind coastal buffer strips. The goose nest is concealed in a 100-yard-wide belt of old growth that is backed by a fresh clearcut. A logging road curves around behind the bear's creek-mouth meadow into what used to be a productive sockeye salmon system. I cannot see the destruction from my camp and only know of it through study of maps and air photos—and conversations with Sylvia Geraghty, longtime resident and guardian of Sea Otter Sound. Industrial-scale logging got a later start here than in rain forests to the south, but it has caught up fast. During the first half of the twentieth century the biggest spruces (the "pumpkins") were plucked from most coastal stands by hand-loggers, yet Prince of Wales remained a vast wilderness of giant forests. Only in the last four decades, under Ketchikan Pulp Company's fifty-year contract, has this island been converted into a 2,200-square-mile checkerboard of clearcuts. Most of the surviving old growth is relatively scrubby. The last truly big timber is the subject of several heated not-in-my-backyard debates.

Today, though, I am on vacation and my only debate is which route to paddle through the enticing snaky passes between the limestone islands of Sea Otter Sound. I choose Marble Passage, a long, narrow channel between Marble and Orr Islands. When the kayak is loaded I slip into the mirror-clear water. The dawn's low fog has melted away, leaving a cloudless blue sky.

At low tide Marble Passage displays the rock for which it was named. The limestone and low-grade marble are cupped and "swiss-cheesy," eaten by sea water. This distinctive topography with its weird internal drainage develops on soluble carbonate bedrock and is called *karst.* Low-elevation karst on Prince of Wales once grew the largest trees in the state. Now, just beyond the narrow old-growth buffer strip, dense five- to thirty-year-old second growth covers the slopes on either side of me.

The story here in Marble Passage is repeated all over Prince of Wales Island. When maps of known carbonate bedrock are overlaid onto maps showing the extent of clearcutting, the coincidence of the two is remarkable and disturbing. Karst is linked to fish, wildlife, and human cultures that depend on high-volume old growth. It also is linked to caves, which on Prince of Wales Island are only beginning to be explored. Here in Marble Passage, I am fifteen miles south of El Cap Pit, at almost 700 feet the deepest-known single-drop cave in the United States. In 1993 the Forest Service, at last acknowledging the quality and extent of karstlands in the Ketchikan district, hired a panel of cave/karst experts, who spent the summer on Prince of Wales and its satellite islands. They found exceptionally productive karst fish streams, a record density of sinkholes, cave-adapted crustaceans that may have survived glaciation in refugia, and archaeological remains including cave art, which is extremely rare in our hemisphere. Perfectly preserved bones of bears and other mammals lying in the caves have recently begun to rewrite the postglacial history of Southeast Alaska and even give hints of life here before the last great ice age. But many of the caves have been destroyed or degraded by logging. And although karst grows huge trees, many karst soils are extremely shallow and all are vulnerable to erosion after clearcutting.

Halfway through Marble Passage I reach a shallow tidal divide, and my hull scrapes barnacles as the current whisks me through the bottleneck. Twice I pass swimming river otters. Their middens up under the trees contain shells of crabs and chitons, eaten in concealment from the pirate-eyed, ubiquitous bald eagles. River otters tirelessly knit the borders of land and sea. They gather their food from fresh- and saltwaters, but they rely

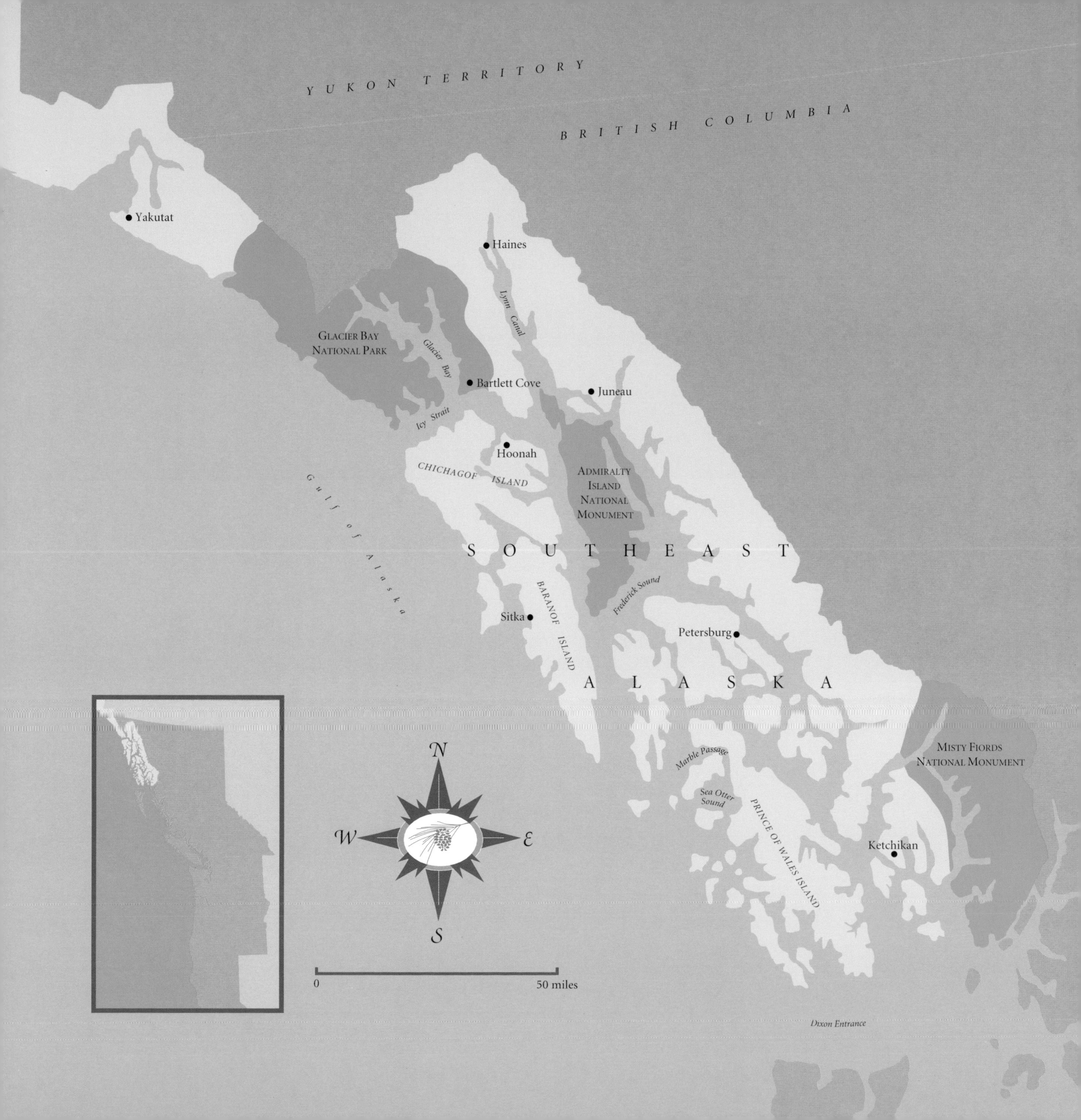

YUKON TERRITORY

BRITISH COLUMBIA

Yakutat

Haines

Lynn Canal

Glacier Bay National Park

Glacier Bay

Bartlett Cove

Juneau

Icy Strait

Hoonah

Chichagof Island

Admiralty Island National Monument

Gulf of Alaska

SOUTHEAST

Frederick Sound

Baranof Island

Sitka

Petersburg

ALASKA

Misty Fiords National Monument

Marble Passage

Sea Otter Sound

Prince of Wales Island

Ketchikan

N

W

E

S

0

50 miles

Dixon Entrance

on the terrestrial forest for rest, cover, and den sites. Otters are probably the ultimate "amphibians," but almost all of Southeast Alaska's mammals partake of the interfingering coastal banquet table, especially in more labyrinthine sections of the archipelago like Sea Otter Sound. My capsule image of this interrelation is a five-foot-diameter Sitka spruce rising 150 feet from the high-tide line with an otter hideaway in its roots, an eagle nest in its crown, and a deer bed in the thick duff of its needle litter. The spruce grows fat on fish fertilizer sprinkled at its feet by the seahunters.

Otters are creatures of the edge, and one might hope they would survive in undiminished numbers as long as a strip of old growth remains along the shoreline. "Not so," says Sylvia Geraghty. "Even the Forest Service anticipates that river otters will fall below viability levels before A.D. 2000 in Sea Otter Sound because of habitat destruction. They den in the forest quite far up the streams, well beyond the narrow coastal buffer fringe that is left visually intact."

Ahead of me, a young buck trots into the water and strokes rapidly for Orr Island, about a quarter mile away. Straining to keep up with him, I can vividly imagine the Tlingit hunters driving deer into the water with dogs, then giving chase in cedar canoes. Near the far bank the buck's feet touch bottom and he splashes ashore. His hooves click and slip on the bare rocks and he trips along splay-legged, then in the sedge turf switches to a graceful lope.

Sitka black-tailed deer range naturally from the latitude of northern Vancouver Island to Lynn Canal at the top of the Southeast Alaska panhandle, and they have been introduced into Glacier Bay, Yakutat, Kodiak Island, and much of Prince William Sound. For twenty years this small cousin of the Columbian black-tail has held center stage in the old growth/logging controversy. That tie to old growth surprises deer watchers in much of North America, where both white- and black-tailed deer initially respond well to the brushy communities that follow logging and have therefore been assumed not to need old growth. But here, at the northern coastal limit for deer, open habitats lie belly-deep in snow through much of the winter, impossible for deer to move around in—and this applies to meadows and peatlands as well as young clearcuts. Only under forest canopy is snow shallow enough to permit easy travel, for multiple layers of overhead branches intercept and hold falling snow. Furthermore, only in *old* forest can deer find enough blueberry browse and low-growing evergreen forage plants to sustain themselves in winter. That is the prelogging, natural reality. Loggers take issue with it, however. They point out that deer are often conspicuously present in clearcuts even in winter, a true observation but a situation that is only phase one. It does not last long. Next come post-logging forests of hemlock and spruce with close-set trunks, dense canopies, and gloomy forest floors. Deprived of light, underbrush in such a second-growth forest dwindles and deer find little browse.

Currently, however, Prince of Wales Island has widespread, recent, brushy clearcuts, and about fifteen consecutive mild winters have kept them only lightly blanketed with snow. Deer have not yet paid the full price for loss of old-growth habitat. The present pace of logging is not sustainable, however, and as it inevitably slows, closed-canopy stands outnumber freshly logged stands. Today's relatively open berry-bush thickets will become impassable conifer tangles with interlacing branches all the way to the ground, too dark and litter-strewn even for mosses to grow. Add a few heavy winters, forcing all deer into remaining patches of old growth and collapsing buffer strips, then also add wolves and human hunters, and skeptics will stop laughing at Fish and Game's predictions. It is a matter of time. Southeast Alaska lacks many of the time-proven old-growth indicator species of the lower forty-eight states—for instance spotted owls and red tree voles—and our vertebrates with apparent connections to high-volume old-growth forest—marten, goshawk, marbled murrelet, brown creeper—are only beginning to be studied. But with deer, researchers since the mid-1970s have gathered reams of data showing that their populations will be drastically reduced by widespread conversion of old growth to even-aged second growth with its rapid cutting rotation.

The wet buck fades into the old growth still lining this shore, and I paddle on, looking for a good campsite on south Marble Island. Today is a short kayaking day because I intend to climb some logging roads to a 1,200-foot ridge overlooking Sea Otter Sound and the open Pacific Ocean.

In the forest, rusty menziesia bushes tower over my head untouched by deer, blueberry stems are browsed and stubbly, and the evergreen fronds of sword fern were clipped last winter. Deep in the shade, under many-layered hemlock branches, I am stopped in my tracks by a rare orchid with waxy basal leaves pressed flat against the step moss: a round-leaved rein orchid, which reaches its northern range limit on Prince of Wales Island. Repeatedly during my travels on this island, I find plants that grow no farther north. Pacific ninebark, bearberry honeysuckle, and salal are novel to me, though so common to the south. As latitude increases from California to Alaska, both plant and animal species gradually drop out, only occasionally replaced by northern interior species like soapberry and nagoonberry and the northern red-backed vole.

This poses a dilemma when trying to characterize Southeast Alaska within the context of the whole sweep of coastal rain forest. Southeast may boast glamorous whales and brown bears and tidewater glaciers, but what about biodiversity? How, I asked Prince of Wales Island wolf biologist Dave Persons, can I get an Oregonian excited about Southeast Alaska if we are only the impoverished tail end of a system whose complexity and productivity peak a thousand miles to the south. Persons was the right one to ask. "It's the archipelago!" he enthused. "What other temperate rain forest in all the world has been diced into such a galaxy of islands?"

On Prince of Wales Island and its myriad smaller neighbors, I have come to share his excitement. No two islands are identical. Their size and isolation from each other shift through time according to the rise and fall of sea level and the sculpting invasions of glacier ice. From island to island, differing complements of predators, herbivores, and seed-dispersing birds result in differing communities. For diversity perhaps we need only to use DNA and skull measurements and look below the species level. The wolf, long-tailed vole, northern flying squirrel, otter, and spruce grouse on Prince of Wales have all been proposed as variants long-isolated from the norms for their species. Island-ness, per se, makes Southeast Alaska and British Columbia's Queen Charlotte Islands a natural laboratory for biogeographers who study the distribution and evolution of organisms. The lessons we can learn may help us predict outcomes of the human-induced isolation to the south, where old-growth "islands" linger amid seas of second growth and human settlement.

Wolves are common on Prince of Wales Island, the mainland, and islands south of Frederick Sound, but not on Admiralty, Baranof, and Chichagof Islands. Only ten of Southeast Alaska's fifty species of land mammals are found on all the major islands; wide marine channels and extensive, rugged mountain barriers inhibit migration.

Having climbed to 100 feet in elevation, I break out onto a gravel logging road. It parallels the coastline for half a mile, then turns upward. Dynamited limestone faces are densely sprinkled with fossilized Silurian-age corals and brachiopods. What a mixing-pot planet we inhabit! Invertebrates of ancient tropical seas, turned to rock, were rafted to boreal latitudes to decorate the walls of bear caves and nourish rain-forest conifers.

Currently wolves seem to be the main travelers on this abandoned logging road, and they evidently are fueled chiefly by venison. Bleached white in the sun, each wolf scat contains deer hair and bone chips. Dave Persons and his colleague Moira Ingles, both of the University of Alaska, radio-collared nineteen wolves on this island. When I talked to them in the spring of 1994, only ten were still alive and most of the others had been trapped or shot. It seems clear that short lives await "disperser wolves" that move from pack ranges in the last of the island's undisturbed wilderness to its new maze of clearcuts and logging roads.

The wolves of Prince of Wales Island have little interchange with those of the adjacent mainland. They have succeeded as an isolated, island population, but they may now be threatened. Unlike their relatives the coyote and fox, wolves cannot survive in highly dissected or "humanized" landscapes, and Prince of Wales road density is already double that at which wolves elsewhere have vanished.

Trappers and others who might be expected to understand the wolf intimately can be deceived by appearances. They point to the abundance of wolves here. Certainly, counters Dave Persons, there are wolves (and deer) in the island's second-growth forest. But these places may be inferior habitats that receive the overflow from crowded remnants of old growth. When we have crossed some as-yet unidentified threshold in the ratio of second growth to old growth, Prince of Wales may prove incapable of supporting viable wolf populations. The last large unlogged block on the island—the Honker Divide—serves a vital role for Dave's wolves. "I can't prove it," he told me, "but logging Honker could mean the end of wolves on Prince of Wales."

My 1985 air photos tell me I should now have reached the top of the uppermost clearcut, but they do not show what happened in 1987. From here to the ridge top 300 feet above me, the steep slope has been stripped of every tree. Naked stumps and slash jut chaotically from chest-high regenerating conifers. Sitka spruce saplings are common, perhaps because soils were heavily disturbed when the precipitous hillside was logged; spruce pioneers disturbance. Leaving the road, I plunge into the prickly tangle. Fortunately, I locate a deer trail, which cleaves straight up the slope, its soils bare and unstable from the constant passage of hooves. I wonder if any of the deer on this ridge are old enough to remember the trails their parents walked through the cool green forest that stood here five years ago.

On the ridge top I sit crosslegged on the forty-inch stump of a Sitka spruce that sprouted around the year A.D. 1600. The sea spreads below me, gentle swells glinting in the setting sun. Two humpback whales spout off Heceta Island, far to the south. Turning, I face the land of clearcuts. In the four centuries it took to grow my sturdy perch, fifteen human generations came and went. Even in the time of our great, great, great, great-grandchildren, Prince of Wales will not regain trees like this.

I remind myself of the richness of the remnants I paddled through today. Euphoria and sadness quit seesawing and melt into a unified resolve.

The Northern Tip of the Rain Forest

There are roughly 70,000 people in Southeast Alaska, the rugged "panhandle" region squeezed between the mountains and the sea and stretching from the British Columbia border near Prince Rupert to Icy Bay, north of Yakutat. Although double the number of aboriginal Tlingit, our entire human population today is smaller than that of Bellevue, the booming city across the lake from Seattle. The Tongass National Forest occupies a large proportion of the panhandle covering some 26,000 square miles, over twice the size of the combined national forests of western Washington and Oregon. If the 70,000 of us were to spread out evenly along Southeast's 15,500 miles of convoluted shoreline, few could see their nearest neighbor. Furthermore, the town-to-forest transition is sudden here—no sprawling belt of burger joints and malls and farms. The typical Southeast Alaskan lives within a five-minute walk of an old-growth forest. From the heart of Juneau's Mendenhall Valley, which is our concession to twentieth-century suburbia, the walk takes only a bit longer—perhaps enough so to warrant riding a bicycle.

Southeast Alaska and rain forests to the south share a maritime climate and many of the same rain-forest plant and animal species. But our northern forests also differ in several ways. The big trees of Southeast Alaska and northern British Columbia are topographically more restricted than the (prelogging!) lowland old growth of Washington and Oregon. At our high latitudes big timber is virtually limited to low elevations and is also associated with certain landform types: flat but well-drained river floodplains; gently sloping alluvial fans where mountain streams emerge onto lowland terraces; ancient uplifted deltas; and steeper but partially stabilized slide areas.

These stands of giant trees are extremely valuable to wildlife and salmon. Since 1950, however, the majority have been logged, and few of the remainders are in protected wilderness. Overall, northern trees are smaller than southern trees of the same species. A century-old Sitka spruce in an average sea-level stand in Oregon will be about 150 feet high, compared with only 100 feet in Southeast Alaska. The rule of thumb is that tree height at 100 years declines northward at the rate of three feet per degree of latitude.

Conversely, although *big* trees are more limited in the north, *old* trees abound owing to relative freedom from certain catastrophic disturbances inherent elsewhere. Here, forest fires are absent except in a few, relatively dry rain-shadow pockets like northern Lynn Canal, whereas periodic wildfire sweeps almost all other natural habitats in North America, occasionally even rain forests as far north as Vancouver Island. Widespread mortality from insects and diseases is also less common in Southeast Alaska than to the south. Epidemics tend to start among trees of similar age, and old growth is characterized by trees of all ages. Another factor is that our wet growing-season weather is hard on tree-infesting insects.

The major disturbances to forests in Southeast Alaska are blowdowns. These storms range in scale from knocking down individual trees to wiping out entire mountainside forests. Small-scale blowdowns perpetuate the uneven-aged structure of old-growth forests. Large-scale, severe storms may take down everything except small subcanopy hemlocks, which then grow into even-aged stands. These, like stands that follow logging, are easily detectable from the air because of their smooth canopies, uninterrupted by the gaps and the uneven tree heights that characterize old growth. At any scale, wind effects

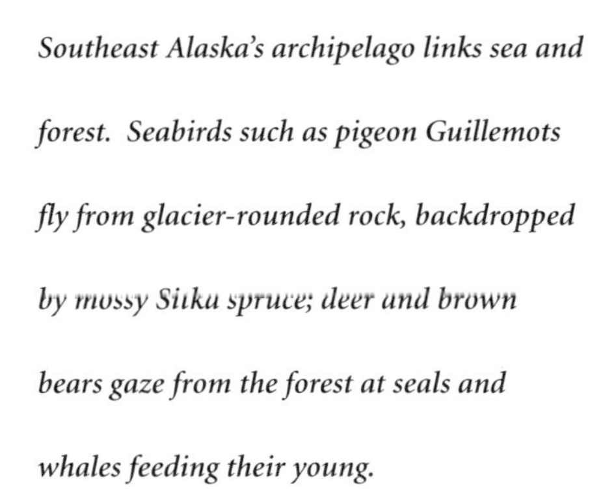

Southeast Alaska's archipelago links sea and forest. Seabirds such as pigeon Guillemots fly from glacier-rounded rock, backdropped by mossy Sitka spruce; deer and brown bears gaze from the forest at seals and whales feeding their young.

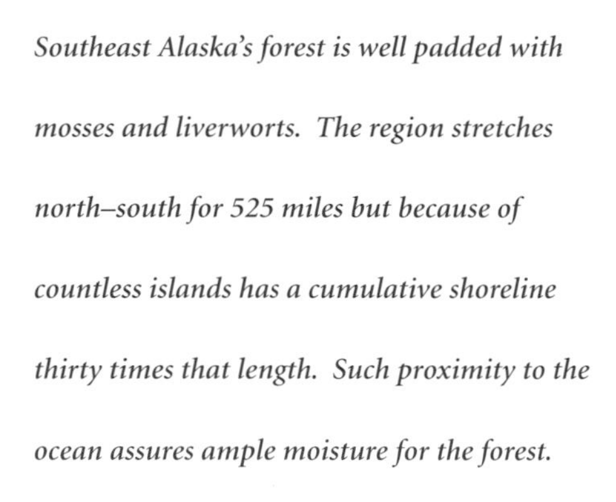

Southeast Alaska's forest is well padded with mosses and liverworts. The region stretches north–south for 525 miles but because of countless islands has a cumulative shoreline thirty times that length. Such proximity to the ocean assures ample moisture for the forest.

vary by tree species. For example, aging western hemlock almost always develop heart rot and therefore tend to snap rather than to uproot in storms, whereas south of here decay-resistant Douglas-fir are more likely to uproot than snap.

Another distinction is that here in the north we have fewer tree species than in the rest of the coastal forest. Pacific yew and Pacific silver fir barely make it into southern Southeast Alaska. Western redcedar reaches Petersburg, perhaps limited to that latitude and southward by growing-season temperatures and winter snow damage to leaders. Shore pine grows in peatland as far north as Yakutat. Western hemlock and yellow-cedar extend beyond Southeast Alaska to Prince William Sound. Sitka spruce is more confined than other rain-forest trees to the narrow coastal belt, yet it persists farther northward than any of the others; it thrives as far as Cook Inlet, where it hybridizes with the white spruce of the vast boreal interior.

The number of shrub and herb species likewise decreases to the north. Their progressive decline can even be observed between southern and northern Southeast Alaska: bugleweed, hardhack, and pygmy water lily gradually drop out. Forest bird and mammal diversity declines too, and the species that remain tend to be hardy generalists. Few are true old-growth specialists, but because our winters are relatively severe, a greater proportion of them depend on old-growth forests during hard times. For these species, open communities like peatland, meadow, thicket, beach, and alpine tundra tend to be fair-weather-only habitats.

In general, the northern forest appears less diverse than those to the south, but there are at least two ways in which we buck the trend. One is our impressive list of bryophytes, the ubiquitous mosses and liverworts; their numbers increase between coastal northern California and Southeast Alaska. The other way is the fine-grained diversity of our habitat mosaic: Southeast Alaska's virtually uninterrupted summer rain assures that freshwater wetlands are liberally sprinkled through the forest. Timberline is low, usually between 1,000 and 3,000 feet, depending on avalanche exposure, aspect, snow-creep, and vegetation's response to the retreating snow line. Lush subalpine meadows therefore abound. Within an hour, a bear that tires of harassing marmots in a ridge-top meadow can be fishing in a river bottom, or licking up cloudberries in a lowland sphagnum bog.

Because of our dissecting waterways and endless miles of meandering shoreline, no point on land is far from the beach. This vastly increases the importance of the ocean and intertidal communities for creatures we typically think of as terrestrial. In actuality the beach for them is highway, pantry, sunbathing spot, bathroom, message board, and birthplace. It is a basic part of life for Southeast people too. Many own more boats than cars, and the twenty-six settlements here with a hundred or more year-round residents all face the ocean, except for Klukwan. All Southeast forests are coastal in the sense that their inhabitants make seasonal or daily use of marine resources.

Partly for this reason, human subsistence directly from wild land still works. Venison, fish, marine and intertidal invertebrates, seaweed, wild greens, mushrooms, and berries are major parts of our diet, especially in small communities. In fact, many local people object to the term "subsistence," which connotes barely surviving. Anyone who lives on coho salmon and deer is royalty, regardless of income. In addition, the most vigorous commercial salmon fishery of the Pacific Coast is here. With low human population and the relatively recent onset of industrial-scale logging, most of our anadromous streams are still healthy and productive, an almost intact legacy. Coho depend on big, old trees. So do deer.

Another factor that sets us apart is that we still have tidewater glaciers, the only active glaciation within the lowland forest from California northward. The northern part of Southeast Alaska is still emerging from the Little Ice Age, which was much more pronounced here than to the south. Our lingering seaward glaciers are so accessible that they began supporting tourism a century ago—and tourism now helps to reduce economic dependence on logging. Ancient forests are impressive in their own right, but ancient forests next to glaciers and viewable from vessels the size of small towns boggle the mind. They interest even the Chamber of Commerce.

A Wild and Perfect Solitude: Admiralty Island

An otter could sniff this track and understand. But I am just a man, so I squat alongside and speculate. I pull out my five-inch comb and lay it on the beach sand for scale, gathering details for tonight's field notes. The front foot measures seven inches across. I wonder if this is the big mother that bluffed my friends off the mountain last summer. The idea spices my evening stroll.

Admiralty Island, to the native Tlingit people, is *Kootznoowoo,* fortress of the brown bear. I concede. She owns the beach and all behind it. Seven-inch tracks—and no others—crisscross the drift line throughout the two miles I have just walked. Frequent half-gallon plops of mashed cow parsnip seed show that she is resident, not just passing through. Departing a little stream full of spawning humpback salmon, she discharged a ten-foot-long dribble of black fish-scat sprinkled with blueberries. No question about which way she was going; the scat begins as a connected stream and peters out into cup-sized splashes. I laugh admiringly. Where else but on this last wild edge of the continent can I trail in the gross wake of giants? As I walk, I pick up, fondle, and discard signs of their passing. The third right-wing feather of an eagle is twenty inches long and has a shaft three eighths of an inch in diameter. It is best held swordlike in the palm, not pen-like in the fingers. Soon, I exchange the feather for a sea lion pelvis. Then I tire of the bone and hang it on the battered root of a drift log. Nobody will see it. People do not walk this beach.

Signs are often all I see of the forest's inhabitants on my visits to Admiralty Island, but they are guarantees of these others' presence. They are infused with stories:

Mid-May, 1,500 feet above sea level. The brown bear sow grazes languidly in the melt water gullies on the edge of a wide, south-facing avalanche chute: she nips sedge tips and sprouts of twisted-stalk, a common lily. Bowed-down stems of Sitka alder periodically rustle, then spring from the snowpack in the midday heat, briefly startling her. Most of her kind are down near the beaches, where forage is more lush and supplemented by the carcasses of winter-starved deer. But this spring she has a new cub, no larger than a beagle, and she shuns the company of cannibal male bears. Besides, she is still groggy, only a week emerged from the den. On the snowy north slope, its door frame an arching mountain hemlock root, the den is half her life. Just now it is still the stronger half. With the cub veering in tow she retreats to the cool montane forest, looking for a place to nap.

Her sleepy reclusiveness is temporary. She is the oldest, heaviest, and most assertive female on this part of Admiralty Island. Last August, fattened and feisty on fish and berries, she and her burly two-year-old son followed a pair of deer hunters for a mile along the ridge top. When the hunters paused to rest, she stood on a boulder close above them, and displayed her gorgeous brown coat with its saddle of silvertips. The hunters clapped and yelled, and politely departed.

Brown bear bravado was commoner a century ago. Sport hunters given easy access by logging roads and kills made supposedly in self-defense are slowly removing the genes and training that produced aggressive bears. Even bear dignity takes a backseat to survival. Part of me sighs with relief; another part regrets the guns that teach 500-pound beasts to whirl and crash into the brush at the first scent of us. All of me hopes that deep in the unlogged mountains of Admiralty, the "real bears" (which First Americans call "Griz") will stand their ground.

Early August, forest above the drift line. I wake from my nap at 7:30 P.M. with a fever and a foreboding tickle in my sinuses—excuse enough for sleeping away the day, if excuses are needed on the island. The "Fortress of the Bears" should understand my drowsiness. Rain patters on the tent fly, and while it is calm here under tall spruces, wind thrashes the alder screen at the beach fringe. Tonight's home is secure from storm but not from furry neighbors. Who knows? Maybe insecurity is itself a need as profound as the craving for food and shelter. Part of each city dweller longs for the nearness of beasts. On Admiralty, all needs met, I huddle in the deepening dusk. I have not eaten but am fed by this wild and perfect solitude.

Scrutinizing that claim of perfection—it seems slightly boastful—I step from the tent wearing nothing but sandals and walk down to the kayak. Crouching under the dripping branches at forest's edge, I discover I cannot see the island I paddled from yesterday. White fog presses low over the three-mile saltwater crossing. Small waves drive northwest. Rain-starved touch-me-nots guzzle a long-withheld drink. Seabeach sandwort on the path to my tent lies crushed by bears' feet. Still crouching, I irrigate the duff. We try to warn each other.

I come back to the tent before the shivers set in and decide to let the claim stand. A wild and perfect solitude.

Mid-September, 2,500 feet. The blue grouse hen freezes, alerted by movement in the nearby elfinwood. A month ago she would have clucked a warning to her two surviving chicks, but they have grown to nearly adult size, and her motherly concern is fading. One chick, oblivious, continues to stuff its crop with over-ripe dwarf blueberries until a less-than-stealthy goshawk fledgling no older than it is glides swiftly over the meadow and sinks her talons in its back. Bowled over, the young grouse flaps violently and tumbles free, but not in time to fly. The goshawk recovers first and pounces again. Disoriented in the swirling cloud of its own downy contour feathers, the chick struggles more feebly and becomes the sloppily dispatched meal of a queen predator-in-training.

Even just weeks out of the nest, the goshawk's strength and agility are astounding. Short, wide wings and long tail permit lightning right-angle turns within forest cover. Her acceleration is instantaneous compared to that of eagles or red-tailed hawks. But unlike red-tails, goshawks are declining throughout their North American range, due to fragmentation of old-growth forests. This female was lucky enough to be born in a nest

Brown bears are also known as grizzlies, their common name in the interior. They are omnivores: both predator and forager. With abundant salmon and a long season of sprouts, berries, and roots, coastal bears fare better than inland bears and are larger. A big male may weigh a half ton and live about twenty years. Note the cub at right.

on Admiralty Island. Alaskan goshawks range widely over huge tracts of ancient forest and its fringing beaches and subalpine meadows. Nearer to towns and homesteads, the autumn training hunts of newly fledged birds are often conducted in domestic chicken coops, where the lessons are ended by shotgun.

Waters in Glacier Bay now plied by cruise ships, kayaks, and fishing boats were blocked by glacier ice when the British mariner Captain George Vancouver arrived in 1792. Today's fish, forests, mammals, and birds are newcomers to both the waters and the land.

At home in Juneau, combined urban distractions and plain laziness make it all too easy to forget one's biological inheritance. On Admiralty, my friends and I seem to become wilder. We go there as some go to church, to be reminded.

The sanctity of the island was finally acknowledged publicly in 1980 when Congress created the Admiralty Island National Monument/Kootznoowoo Wilderness, to be managed by the U.S. Forest Service. That preservation was achieved not by, but in spite of, its present federal guardian. Admiralty Island had several times been saved from logging under mercifully aborted long-term contracts, and to assure its future, an unlikely coalition came together: grassroots environmental groups, hand-trollers, native subsistence users, sportsmen's organizations, bear guides, and "outside instigators" like the Sierra Club. Until designation of the wilderness, no large blocks of productive lowland old growth in Tongass National Forest had any formal protection.

Nearby Glacier Bay National Park is very much a travel destination (visited by a quarter million out-of-staters annually), but 100-mile-long Admiralty Island has no jet or tour ship landings and few lodges or developed facilities. Except for bear-watching tourists at Pack Creek, who shuttle to and from Juneau in floatplanes, the island remains primarily the subsistence preserve of Southeast Alaskans who both savor and rely on its intact legacy. Indeed, Admiralty's importance to surrounding communities increases as Tongass logging degrades once-productive forest lands elsewhere.

Officially, the Admiralty Island National Monument/Kootznoowoo Wilderness is managed as one unit from offices in Juneau. Traditionally, however, the island belonged to not one, but four groups—the Hutsnuwu, Kake, Taku, and Auk—and in practice today it is still divided among the communities that dot its saltwater channels. There still are almost no roads, leaving virtually all access by water. Motorized skiffs may be faster than paddled cedar canoes, but because their owners usually have to be back at work on Monday, their effective range differs little from that of the old days.

Despite this isolation, the island shares wounds common to all Northwest forests. Here, as elsewhere, it is truly a challenge to find a high-volume coastal stand without the most ubiquitous reminder of the hand-logger era: mossy stumps sawed off eight feet or so above the ground, larger in diameter than any tree now remaining. It has become my habit to mentally replace the lost monarchs, reconstructing the forest of 1900, and to wonder what, aside from grace, we have forfeited. But anyone who has traveled real war zones, like Prince of Wales Island, finds Admiralty an island of blessed solace. From my home in Juneau I can look across Stephens Passage to twenty miles of Admiralty's northern hills and see not a single logging scar. To those who stood up for the island, I owe my food, my retreat, perhaps even my sense of self.

In 1986 Admiralty Island and Glacier Bay were jointly added to the distinguished company of world Biosphere Reserves. The two make a disparate but complementary pair—Admiralty's forests hoary with age, tectonically stable; Glacier Bay raw and youthful, in places barely vegetated, its "teenaged" mountains fast-rising from colossal colliding terranes. Both extremes—youth and age—have proven attractive to forest ecologists. Researchers come to Glacier Bay from all over the world, but forest research on Admiralty is conducted primarily by local Alaskans. Glacier Bay changes dramatically from decade to decade; on the island, in places where the forest is sheltered from wind, millennia may pass between major disturbances. The resulting Admiralty Island plant communities offer rare insights into successional trends that cannot be detected within a human lifetime, or even that of an individual tree.

Most old-growth researchers in Alaska have worked on Admiralty. It is reassuring to know that one's study site will probably survive intact well into the coming century. Paul Alaback, the acknowledged authority on Southeast Alaska's old growth (formerly with the Forestry Sciences Laboratory in Juneau), has eleven of his sixty Southeast forest study plots on Admiralty. Bob Deal (also of the Forestry Sciences Lab) has done intensive stand-reconstruction work, hoping to understand, tree by tree, the history of a forest whose elders germinated in the 1400s. Paul Hennon, a colleague, is investigating a mysterious decline in the yellow-cedar that now covers more than a half million acres in Southeast Alaska, "the most spectacular forest decline in western North America." Answer so far: unknown.

Researchers Matt Kirchoff, John Schoen, LaVern Beier, and Kim Titus (all with the Alaska Department of Fish and Game) study how deer and brown bear that depend on old growth are affected by logging, mining, and roads. Recent genetic work on the brown bear suggests those on Admiralty may have been isolated from bears on the adjacent mainland for a long time, an echo of the apparent wolf isolation on Prince of Wales Island. Other researchers working on Admiralty Island (University of Alaska Museum teams directed by Joseph Cook and Stephen MacDonald) have found native muskrats, martens, and meadow voles that are not present on nearby Baranof and Chichagof

Islands. Admiralty's meadow vole and also its beaver are subspecies distinct from those of the adjacent mainland. This distinction suggests prolonged isolation. Probably the Tongass, extraordinarily endowed with islands, has more endemic mammal subspecies than any other national forest.

Recently I sat in a room packed with biologists. The assembled experience and knowledge of Southeast Alaska was exceeded only by a palpable devotion to the land, a mix of love and wisdom I have also felt in the presence of Tlingit elders. Among other considerations, we groped for ways to nurture certain forest-dependent animals—goshawk, murrelet, brown bear—on unprotected parts of Tongass National Forest. Opponents of legislated wilderness describe such protection as an elitist "lock-up." Yet how much more drastically can we lock up land than by felling ancient forests? The Admiralty Island National Monument/Kootznoowoo Wilderness is in fact a last bastion of *accessibility* for the low-income subsistence user, the bear-watcher tourist, and the researcher. For each, Admiralty is sacred.

Glaciers calving icebergs into saltwater—such as Glacier Bay's Lamplugh Glacier—are now vanishing. The ice is melting. Twenty thousand years ago ice overrode most of the Alaska and British Columbia coasts, sculpting a maze of deep, broad valleys. Now flooded by the sea, the valleys are labyrinthine fiords.

The Long Healing: Glacier Bay

In this region of bays there is only one *Bay.* Even for those who have seen no more than the ubiquitous magazine ads and travel guides, "the Bay" conjures images of breaching whales and wave-lapped glaciers. Those images, however dramatic, are understated. Glacier Bay hosts the biological recolonizing fervor of land-rush Oklahoma, framed by rock and ice and so elusive of description that American writers since the time of John Muir spin their wheels, blurting superlatives in the first premature paragraph. I am no different.

Thirty whales and 160 tour ships per year cannot be wrong. Glacier Bay's riches support, for now at least, a simultaneous influx of big-league tourism and big wilderness-only mammals like orcas, wolves, and brown bears, drawn to waters and land newly freed of ice. But Glacier Bay National Park, in its former status as National Monument, was not created to woo recreation dollars. As conceived by the University of Minnesota's pioneering plant ecologist William Cooper, the Bay would serve science. It would allow a succession of researchers to untangle the complexities of a succession of plants, in perpetuity.

Who, or what, is wilderness for? At the first Glacier Bay Science Symposium in 1983, one unsuspecting researcher concluded his talk with the observation that science was the "highest use" of the park, that science—unlike other human activities—was at once non-consumptive and a benefactor to all, from recreationists to extractive users like the

halibut fleet. An environmentalist responded that recreation, in *its* highest form, goes far beyond diversion. In the Bay we tap our legacy and re-create humanity's bond to the Earth, preferably in silent wonder, out of hearing of scientists' floatplanes and outboard motors.

The Hoonah people espouse neither science nor recreation. They mourn a lost home. Across Icy Strait from Glacier Bay, Hoonah village has among its elders those who fished and hunted and gathered in the Bay as children. To the Chookanedi clan, Glacier Bay was never wilderness. Their earth was not divided into outdoor laboratories, untouchable museums, and one-shot resource-extraction areas. Nor was it a playground. This reef was where to hunt seals. That cove was *Tsulk ge geiyi,* a "place with many marmots." This clean sand grows spruce with long straight roots, twenty feet with hardly a taper, to split for basketry. *Gaat heeni tlien* is "big sockeye creek." *Xwat aanyi luee* means "seagull islands are close," for the gathering of eggs, summer after summer.

Dwarf fireweed, horsetail, and willow quickly pioneer bare rubble as glacier ice melts. So do mosquitoes and blackflies! All are curtain-raisers for life's reconquest of the land.

June 27, Upper Glacier Bay, West Arm. Our kayaks and gear lie strewn on the barren beach, hurriedly dumped from the drop-off ship, which has chugged away leaving us in silence. For ten days we will soak in this silence, broken only by our own usually muted voices, the occasional noise of wind and waves, and at times the cries of thousands of birds. We will paddle sixty miles from bare rubble here at bay head to 200-year-old spruce forest at bay mouth, a pace so slow we will almost be able to imagine sitting at one place for two centuries and watching a rain forest spring from earth newly free of glacier ice.

I am here to assist Greg Streveler, trip leader, longtime resident of Gustavus (near the mouth of the bay), and unofficial but widely acknowledged master natural historian of Glacier Bay. His status is so unofficial that our guests will not realize until perhaps halfway down the bay that the guy in the goofy, inside-out-and-backwards sou'wester cap has served for twenty-five years as guide to world authorities in every field from bedrock to bugs. But facts about the environment, however encyclopedic their source, are just so much "nature tour." What sends people home to Ohio refreshed is the daily ritual of travel through wild land, powered by their own bodies, fed in token part by wild fruits and leaves, and guided by someone who loves as much as knows.

In camp, Greg tells how the great British mariner Captain George Vancouver in 1794 sailed past the entry into what is today Glacier Bay. He did not see it. A tidewater glacier eight miles wide was plugging the waterways we will be exploring. Since that time, the ice has retreated more than sixty miles, melting back into scores of separate tributary lobes. Forest growth on the lifeless rubble exposed by the retreat of the ice is called *primary succession*. It is exceedingly unusual on so vast a scale. Most ecology textbooks therefore cite the Glacier Bay succession studies by pioneering ecologists William Cooper and his successor Professor Donald Lawrence.

Floating on milky, placid water our novice kayakers stare expectantly at the dripping face of Lamplugh Glacier. A recently exposed section of the 300-foot ice cliff is surrealistically blue; if we were crazy enough to paddle closer, we could probably see stones embedded ten feet back inside the clear, dense ice. With a warning boom, the wall next to the blue section collapses, and enough ice to cool all the mixed drinks in Alaska for a year slides majestically into the sea. We point our suddenly feeble-looking craft into the wave that rushes toward us, but Greg has positioned us safely and the wave is reduced to a long tame swell by the time it arrives.

Sometimes expedition guests are disappointed that we begin our trips at the world-famous tidewater glaciers, and leave them after only a day or two for less showy realms down-bay. But unless you are a kittiwake plucking churned krill from the icy waters, glaciers calving off icebergs are like a fireworks display, better for epiphany than for daily fare. The climax of Glacier Bay to a naturalist is the mouth, not the head; it is what life has wrought from what ice destroys. We travel through time, from immediately post-glacial toward stately forest, and Greg and I fully expect our companions to become infected with passion for this pageant of changing life. It helps when store-bought greens give out on day three or four. Naked glacial till is then less appealing than the more lush beaches of the middle bay. We study salad succession.

June 29, snout of Reid Glacier. In the first days of the trip, plants are often just lonely sprigs here and there among the rocks. We are so far up-bay from the nearest forest that even the efficient winged seeds of conifers have failed to move in. More successful are the plumed seeds of willow and cottonwood, and the microscopically tiny seeds of wintergreen and spores of ferns. Rock mosses and Easter and dogtooth lichens are common early colonizers. Sometimes we find soapberry seedlings germinating in the desiccated scat of a wide-ranging brown bear, and kinnikinnick is frequent in the seed caches of tundra voles, which are limited by small size to their home territory.

A favorite argument among plant ecologists concerns the mechanisms of succession, with its interface between competition and facilitation. For example, do some pioneering plants "help" those that follow them, or do they inhibit them? Both effects are evident in Glacier Bay, according to early studies by Professor Lawrence (University of

The forest at Bartlett Cove has developed since the retreat of glacier ice. It appears timeless, but nature inevitably follows a dynamic course. Insect infestation may create gaps, and soils may turn soggy because of blocked drainage. Roots will waterlog, trees topple, and acid conditions increase. Today's forest may become tomorrow's open peatland.

Minnesota) and recent studies by Terry Chapin (University of California, Berkeley) and Lars Walker (University of Nevada). Beyond question, nitrogen-fixing species like mountain avens and Sitka alder enrich the soil; they help. But their presence may actually slow willow and cottonwood; they compete. The idea of sequences from species that prepare the way to species that follow is orderly, but the reality is complicated.

July 3, Drake Island, middle bay. We have played with the concepts of how green life recarpets deglaciated land throughout the first half of our ten-day journey—barren ground grading to thickets, and thickets becoming overtopped by fifty-foot cottonwoods. Next come the mixed Sitka spruce–black cottonwood forests of the middle bay, my favorite part of the trip.

By this stage, guests have stopped looking wistfully over their shoulders toward the stark blue-ice glacier country behind us; there are too many attractions right here. A moose herds her calf into the willows. An eagle glares from her nest in a beachside cottonwood. Humpbacked whales blow explosively. At every beach stop the density of tracks increases: wolf, coyote, otter, mink, moose, goose, oystercatcher. Greg draws in the mud, showing us how to distinguish black from brown bear prints; blackies have a more arching and widespread toe configuration.

At about 120 years since retreat of the ice, these forests are very different from post-logging stands of similar age. Instead of logging's legacy—a single-layered, interlocking conifer canopy and an impoverished moss-dominated understory—the mixed coniferous/deciduous forests of middle Glacier Bay offer optimum habitat for many vertebrate species. Spruce trees grow far apart, with long, sturdy branches that provide cover for winter residents. Ninety-foot cottonwood with a subcanopy of Sitka alder and felt-leafed willow assure nesting songbirds an abundance of leaf-eating insects. Bird diversity peaks, and berry-producing shrubs are far more common here in mid-bay than in the tangle of alder that precedes them or in the closed-canopy spruce forests of the lower bay. As we complete a map of a sixty-foot-square plot, Streveler points out that each of the nine subplots within it is distinctive, both in canopy structure and in understory species. By northern standards this is a remarkably diverse forest.

The bare tree trunks of a forest buried by glacial gravel thousands of years ago are now eroding free. They act as a marker within the sediments' great layer cake of time. They are a reminder of nature's constant change.

July 5, Beardslee Islands. As our kayaks near the lower bay, deglaciated for nearly two centuries, we camp among simpler, even-aged forests. Sitka spruce predominates. Until recently hikers delighted in these forests; it was easy to stroll in any direction over park-like, brush-free, moss-carpeted moraines. But then the forests abruptly began to change character. The spruce started dying.

No ecologist foresaw this. Spruce forest usually is a protracted stage. Yet here in the lower bay, spruce bark beetles—bullet-shaped tree girdlers a quarter-inch long—infested more than 10,000 forested acres, killing 40 to 70 percent of the dominant trees. Such vast invasions are rare in Southeast Alaska. Only in extraordinary places like the recently deglaciated lowlands of Glacier Bay are there truly vast, same-aged stands of conifers well past their peak annual growth rate and therefore declining in vigor. In retrospect, we can see these conditions were virtually courting an epidemic. When it happened, Professor Lawrence immediately recognized the importance of this new twist in the Glacier Bay story. He encouraged the establishment of one hundred permanent study plots in the Beardslee Islands and at Bartlett Cove. Since 1984, biologist Mark Noble and I have followed the fortunes of these small, scattered plots as trees die and fall around them, and as lady fern, devil's club, red elderberry, and hemlock saplings spring into the light.

Backdropped by the 15,000-foot Fairweather Range, the Beardslee Islands are unique as well as beautiful. No longer weighed down by overlying glacier ice, they are rising from the sea as much as an inch per year. Some scientists think this rebound may be faster than any other on earth.

The infestation seems to have largely run its course, and surviving trees are growing better than before the great thinning-by-beetle process occurred. But getting from one plot to another involves acrobatics as much as hiking; we sometimes teeter along for hundreds of feet atop fallen logs that are supported six feet off the ground by undecayed branches. Rarely are all the original spruce killed, and survivors large and small flourish where infestation-caused openings in the canopy have increased the light reaching the forest floor. It is simplistic, but tempting, to call the resulting forest instant old growth. It lacks antiquity but has suddenly gained many of old growth's structural attributes: abundant dead wood, both standing and down; a legion of early-phase decomposing organisms; rich, multileveled canopies with perhaps overly spacious gaps; and an incredibly fruity shrub layer. Usually old-growth forest structure in Southeast Alaska takes about 400 years to develop, but in Glacier Bay beetles seem to be hastening the process.

I once stood with about twenty scientists and naturalists in a clearing in such a beetle-opened forest. Greg Streveler led the group through a brainstorming session, drawing on each researcher's specialty: What did this forest look like a century ago? What will it become a century hence? What set this forest up—like a well-placed volley-ball—for the spike that came in the form of a minuscule but mighty beetle? I found myself imagining the swaths of even-aged conifer stands that must have stretched across the continent at the waning of the last great ice age. What lucky insect discovered *those* feasts?

Glacier Bay will always surprise us, both with new events and by demanding reinterpretation. Ecologist Chris Fastie recently studied forest development throughout the bay,

using stand reconstruction methods. He finds that the succession of one type of vegetation by another is anything but uniform from place to place. The popular assumption had been that traveling down-bay from the sterility of a calving glacier to the mossy spruce forest of the lower bay was equivalent, for tracing vegetation sequences, to a humanly impossible stay of two centuries at Bartlett Cove, watching the forest develop. But this often-quoted concept neglects differences in seed sources and whether surface sediments available to plants are coarse or fine—and all the linked events that flow from these two initial conditions. In actuality, succession follows many patterns.

At Bartlett Cove spruce neither competed with, nor benefited from, alder thickets during the forest's first century. The spruce grew slowly and close-set for their first hundred years; in time, became a banquet for beetles; and in places have also experienced a soil change from being well drained to developing a hardpan layer and therefore the beginning of a shift from forest to bog. In contrast, at Muir Point, three kayak days up-bay from Bartlett Cove, fast-growing, century-old spruce is succeeding alder; we saw their dark green spires rising above dense alder thickets. At our first, upper-bay camps—so recently deglaciated—future plant sequences quite surely will be still different. The mountainsides there have no ready source of seed for either alder or spruce.

July 7, last camp in the Beardslee Islands. As kayakers near the end of a long journey they often fall prey to a derangement known among guides as "goin' fer the barn." On our last night out we sometimes take preventative measures, perhaps sitting in a circle, each saying frankly what the trip has meant to us. I remember one man who was so moved that he kept talking, confessing, right through the wolf song that suddenly rose from the forest behind him, bearing its own, wilder witness. A duet.

Why do we need wild land? For science? Recreation? Sacrament? Subsistence? Is wilderness an investment for our children? A legacy from our forebears? Maybe beneath all of our factious squabbling lies striving for clearer, wiser vision. Maybe in officially proclaiming that a certain piece of land cannot be bought, we strain to recognize—and restore—the dignity of the nonhuman world. By legislating wilderness we are saying that the unownable future has value.

And what, ecologically, is ancient forest "for?" Maybe it is much the same—the future. A young system, like a meadow or a salt marsh, is nature's breadbasket giving to neighboring communities much of what it produces each summer. An old-growth forest is nature's bank, seasonally more stable, offering less lavish summer growth but adding winter refuge and greater structural complexity. Ecologists try to quantify community productivity, expressing it as P/B, where P is annual production, or the weight of plant material grown yearly, and B is total community biomass, or weight of organic material sustained from year to year. By this standard a young, fast-growing forest is far more productive than an ancient one. But can we contrive a formula that acknowledges the value of community age? Ecologist Raymond Margalef suggested in the 1960s that we might turn the fraction around—B/P instead of P/B—to describe something our Gross-National-Product-oriented society is not used to valuing, or even considering. This view perceives biomass as more than weight, seeing it as lives and systems and cycles, the entangled home of countless seen and unseen interactions.

How much dead wood and canopy complexity, and how many spider webs and murrelet feces, as well as, yes, how many board feet, including that riddled with "defect," can a given acre support for centuries on end without yielding it to wind, tide, ice—or chain saw? Some ecologists think of this strange, unharvestable value as the influence that a community has on its own future. Maybe the real essence of the concept survives only in teasing shreds from a time before we imagined that a forest could be owned and instead recognized ourselves as part of the whole.

Acknowledgments

Seven of us have worked on this book and received crucial help and encouragement from colleagues, friends, spouses, relatives—and The Mountaineers Books staff.

John Sawyer particularly acknowledges his wife Jane Cole.

Kathie Durbin thanks Chris Maser, Michael Donnelly, Tim Lillebo, Kevin Scribner, Barbara Ullian, and Wendell Wood for sharing the secrets of some of their favorite old-growth forests.

Tim McNulty is especially grateful to Jim Hidy and Don Williamson for their help in understanding the singular nature of the Long Island cedar grove; Dr. Nalini Nadkarni, Hugh Mortensen, and Dr. David Shaw for sharing expertise regarding old-growth canopy ecology; Dr. Peter Frenzen for insights into the processes of forest recovery at Mount Saint Helens; Jan Henderson and Robin Lesher for tolerating endless questions during a field trip; and Dr. Peter Morrison for guidance into the complex forest ecology east of the Cascades. For help and overall encouragement, he also thanks Bruce Moorhead, Mark Lawler, Charlie Raines, Susan Saul, and his wife Mary Morgan.

Yorke Edwards is indebted to Dr. Vladimir Krajina for introducing him to the ecology of western forests; Dr. Robert Ogilvie for providing current data on rare plants; David Nagorsen, Stan Orchard, and Grant Keddie for sharing information on the latest research on mammals, amphibians, and archaeology; and his wife Joan for accepting those hours when body is present but mind is far into planning future paragraphs.

Richard Carstensen thanks Cathy Pohl and Greg Streveler for reviews, advice, and years of encouragement; Dr. Paul Alaback, Matt Kirchoff, and Sylvia Geraghty for discussions about Alaskan old growth; and Dave Persons and Moira Ingles for information about their wolf research.

Charles Mauzy is especially grateful to his wife Linda for her support; his daughter Ariel for help with reference material for the maps in this book; his father Byron Mauzy for ongoing support and encouragement; and his colleagues David and Bonnie Muench for creating many new images of Northwest forests specifically for these pages.

Ruth Kirk gratefully acknowledges the five writers of this book for their diligence and dedication and also those others who helped shape the manuscript. Dr. Jerry Franklin contributed greatly to the overall structure of the book; Drs. Dick Behan and Bill Ferrell read the entire manuscript and offered many useful criticisms; Dr. Bill Calder helped sort out a particularly bothersome section; Dr. Nalini Nadkarni checked the section on lichens; Dr. Richard Daugherty and Wayne Kirk gave helpful input on overall coverage and clarity.

To all, the thanks of us all.

Western red cedar

Glossary

Biological diversity (often shortened to biodiversity): the variety of life and its processes at all levels of organization.
Biomass: total mass of all organisms within a unit of space at any given time.
Bog: a peatland in which the water table is at or near the surface and precipitation is the major source of water; trees may or may not be present.
Canopy: the uppermost layer of a forest; the treetops and the structure created by their combined upper trunks, branches, twigs, and leaves.
Climax growth: the culmination of plant succession; a self-perpetuating equilibrium that will continue unless the environment changes; a concept now questioned because change is constant.
Coarse woody debris: logs, fallen branches, and stumps, whether sound or rotting, which provide habitat for fungi, plants, animals, and other organisms.
Composition (of a forest): the particular species of plants, animals, and other organisms that make up a biological community.
Disturbance: biotic or abiotic forces that disrupt growth; typical examples are floods, fires, mudflows, blowdowns, and insect infestations.
Ecosystem: a basic, functional unit of nature that encompasses both organisms and their environments; ecosystems may be of any size.
Endangered species: species or populations likely to become extinct.
Endemic: a species that is unique to a specific locality, often as a result of having survived a widespread disturbance such as glaciation.
Health (of an ecosystem): the ability to maintain complexity and self-organization; the capacity of the species and associations characterizing an ecosystem to remain intact.
Hyphae: the threadlike structures that constitute the body of a fungus.
Hyporheic zone: the water-saturated sediments under a riverbed and floodplain, which are inhabited by a host of little-known organisms.
Karst: a topography characterized by caves and sinkholes, which results from the action of water on soluble rock such as limestone, marble, or dolomite.
Lichen: the merging of algae and fungi (and often also cyanobacteria) into an independent life-form that is distinct from its components.
Litterfall: the organic debris and accumulated compounds and minerals that rain down through the crown of a tree or shrub.
Mature: refers to the growth stage at which the mean annual production of an individual tree or a community has peaked.
Muskeg: a word of the boreal-forest Cree Indians of Canada, meaning an extensive boggy area.
Mycelia: collective term for fungal hyphae.
Mycorrhiza: the symbiotic relationship between trees and fungi, which results in tree roots being better able to take up soil nutrients.
Nitrogen fixation: the conversion of nitrogen into a form usable by plants.
Nurse log: a fallen tree that acts as a habitat for mosses and seedlings.
Old growth: a forest that has passed maturity and is made up of living and dead trees; what constitutes old growth varies by forest type and biogeoclimatic zone.
Peatland: an old wetland with saturated, undecayed remains of plants.
Pleistocene: the first epoch of the Quaternary Period from 2 million to 10,000 years ago; the most recent period of major glaciation.
Rain forest (of the temperate zone, as distinguished from the tropics): a forest characterized by abundant flowing water; acid soils; layers of growth that grade into one another; abundant epiphytes and mosses; great amounts of coarse woody debris; and trees that include the largest and longest-lived of their kind.
Refugia: habitats that support organisms in an area that represents a small fragment of their previous geographic range, such as "islands" of land surrounded by glaciers.
Riparian area: land adjacent to a stream, river, lake, or pond and influenced by that proximity.
Salvage logging: the harvesting of damaged timber; often carried out after a fire, insect attack, or blowdown.
Second growth: regrowth following drastic disturbance, whether natural or the result of logging.
Silviculture: the discipline of producing and tending a forest.
Snag: a standing dead tree or part of a dead tree.
Stand: trees that are uniform enough in species and age to be distinguishable as a group from the adjoining forest.
Stolon: an underground branch, or "runner," that develops a root at the tip and then produces above-ground growth, which becomes an independent plant.
Structure: the physical elements of a forest and how they are arranged; forest "architecture," per se, without regard to species or functions.
Succession: successive stages of plant growth from bare soil to climax.
Sustainability: a balanced relation between healthy ecosystems and human need for their resources.
Taxa (plural of taxon): refers to species and subspecies.
Threatened species: plant or animal species likely to become endangered within the foreseeable future.
Truffles: fungi that live entirely underground.
Ungulate: the taxonomic order of mammals with hooves.

Suggested Reading

Agee, James. *Fire Ecology of Pacific Northwest Forests.* Washington, D.C., and Covelo, Calif.: Island Press, 1993. Discussion of the role fire plays in Northwest forests.

Arno, Stephen, and Ramona Hammerly. *Northwest Trees.* Seattle: The Mountaineers, 1977. Outstanding popular discussion of common tree species; wonderful drawings.

Alaska Geographic. *Admiralty Island: Fortress of the Bears.* Anchorage: Alaska Geographic, 1991. A loving view of the island.

Alaska Geographic. *Southeast Alaska.* Anchorage: Alaska Geographic, 1993. Scientific and geographic portrait.

Dietrich, William. *The Final Forest: The Battle for the Last Great Trees of the Pacific Northwest.* New York: Simon & Schuster, 1992. Views ranging from those of loggers, millworkers, and timber company executives to researchers, environmentalists, and politicians.

Franklin, Jerry F., et al. *Ecological Characteristics of Old-Growth Douglas-Fir Forests.* General Technical Report PNW-118. Portland: USDA Forest Service, Pacific Northwest Forest and Range Experiment Station, 1981. Major ecological features of coniferous forests in the Douglas-fir region.

Hawken, Paul. *The Ecology of Commerce.* New York: HarperCollins, 1993. Analysis of capitalism from an ecological perspective.

Henderson, Jan A., David Peter, Robin Lesher, and David Shaw. *Forested Plant Associations of the Olympic National Forest.* R6 Technical Paper 001-88. Portland: USDA Forest Service, Pacific Northwest Region, 1989. Plant associations and forest types together with geology, fire history, climate, etc.

Johnston, Verna R. *California Forests and Woodlands: A Natural History.* Berkeley: University of California Press, 1994. The diverse life of California's forests.

Kirk, Ruth, with Jerry Franklin. *The Olympic Rain Forest, an Ecological Web.* Seattle: University of Washington Press, 1992. Ecology of our most magnificent temperate rain forest.

Langston, Nancy. *Forest Dreams, Forest Nightmares: The Paradox of Old Growth in the Inland West.* Seattle: University of Washington Press, 1995. Ecology and history in the Blue Mountains.

Maser, Chris, et al. *The Seen and Unseen World of the Fallen Tree.* General Technical Report PNW-164. Washington, D.C.: USFS/BLM, 1984. Discusses the vital role of dead wood in the forest community.

McNulty, Tim, and Pat O'Hara. *Washington's Wild Rivers, the Unfinished Work.* Seattle: The Mountaineers, 1990. River ecosystems, including riparian forest communities.

Meidlinger, Del, and Jim Pojar, eds. *Ecosystems of British Columbia.* Victoria, B.C.: Ministry of Forests, 1991. Forest description covering the ecological regions of the province.

Moir, William H. *The Forests of Mount Rainier National Park: A Natural History.* Seattle: Northwest Interpretive Association, 1989. Examination of Mount Rainier's splendid forest written by a savvy forest ecologist.

Nelson, R. *The Island Within.* New York: Vintage Books, 1989. Sensitive and informative examination of an unnamed Southeast Alaska island.

Norse, Elliott A. *Ancient Forests of the Pacific Northwest.* Washington, D.C., and Covelo, Calif.: Island Press, 1990. Old-growth forest ecology written for the general reader.

O'Clair, Rita M., Robert H. Armstong, and Richard Carstensen. *The Nature of Southwest Alaska: A Guide to Plants, Animals, and Habitats.* Bothell, Wash.: Alaska Northwest Books, 1992. Habitats and ecology of Southeast Alaska as well as its creatures, plants, and plantlike organisms.

Perlin, John. *A Forest Journey: The Role of Wood in the Development of Civilization.* New York: W. W. Norton & Company, 1989. Excellent account of rise and fall of civilizations as mirrored by exploitation of their forests; Mesopotamia to the American frontier.

Schrepfer, Susan R. *The Fight to Save the Redwoods: A History of Environmental Reform, 1917–1918.* Madison: University of Wisconsin Press, 1983. Excellent portrayal of a long, diligent effort.

Sedell, James, and Chris Maser. *From the Forest to the Sea: The Ecology of Wood in Rivers, Estuaries, and Oceans.* Delray Beach, Fla.: St. Lucie Press, 1993. Study of the long-term value of coarse woody debris.

Seideman, David. *Showdown at Opal Creek.* New York: Carroll and Graf Publishers, 1993. Cultural and economic clashes in the North Santiam Valley.

Wallace, David Raines, *The Klamath Knot: Explorations of Myth and Evolution.* San Francisco: Sierra Club, 1983. Evocative essays on the geology, flora, and fauna of the Klamath mountains.

Webb, Clinton. *The Status of Vancouver Island's Threatened Old-Growth Forests.* Vancouver, B.C.: Western Canada Wilderness Committee, 1992. Surveys the island's remaining old forests.

Wood, Wendell. *A Walking Guide to Oregon's Ancient Forests.* Portland: Oregon Natural Resources Council, 1991. Where to find the best old-growth stands.

Index

About the Editors

Ruth Kirk grew up on a hill in prefreeway Los Angeles when vacant lots still offered horned toads to play with, trees to climb, and wildflowers to pick—an outdoor life for a city girl.

With her park ranger husband Louis (now deceased) she moved to Washington from the southwest desert in the early 1950s, living first at Mount Rainier, later at Olympic National Park. Trips doing field research and photography for books and, by the 1970s, films for television documentaries, took the Kirks repeatedly through the forests of Washington, Oregon, and northern California and to the remote corners of Alaska and British Columbia.

She is the author of numerous books, including *The Olympic Rain Forest: An Ecological Web* (with Jerry Franklin); *Exploring Washington's Past: A Road Guide to History* (with Carmela Alexander); and *Desert: The American Southwest,* which was nominated for a National Book Award and won the John Burroughs medal for nature writing.

Charles Mauzy has been a professional nature photographer for twenty years. He is best known for his photographs of Pacific Northwest forests and mountains. His work has been published internationally and his magazine credits include *National Geographic, Life, Time, Newsweek, Sierra, Audubon, Smithsonian,* and *Pacific Northwest.* His photographs have also been featured in numerous advertising campaigns and annual reports for Fortune 500 companies including Boeing, Kodak, Nike, Ford, and Weyerhaeuser

He was formerly Director of Photography for Allstock, one of the largest stock agencies in the country, and he is currently Director of Media Development for Corbis Corporation in Bellevue, Washington.

Founded in 1906, **The Mountaineers** is a Seattle-based nonprofit outdoor activity and conservation club with 15,000 members, whose mission is "to explore, study, preserve, and enjoy the natural beauty of the outdoors. . . ." The club sponsors many classes and year-round outdoor activities in the Pacific Northwest and supports environmental causes by sponsoring legislation and presenting educational programs. The Mountaineers Books supports the club's mission by publishing travel and natural history guides, instructional texts, and works on conservation and history.

For information write or call The Mountaineers, 300 Third Avenue West, Seattle, WA 98119; (206) 284-6310.

The Mountaineers Foundation, a public foundation organized exclusively for charitable, scientific, literary, and educational purposes, is a separate corporation from The Mountaineers and has its own board of officers and trustees. Since 1968, the Foundation has funded projects that promote the study of the mountains, forests, and streams of the Pacific Northwest and that contribute to the preservation of its natural beauty and ecological integrity.

One of the major responsibilities of the Foundation is the preservation and protection of more than 200 acres of old-growth forest and additional buffer property of second growth, west of Bremerton, Washington.

Foundation-supported projects have been conducted by a variety of environmental and conservation organizations. Typical examples include a study aimed at biodiversity management of watersheds; the purchase of land to protect the habitat of bald eagles and nesting migratory birds; the development of an environmentally responsible wilderness education program for young people, and the development of a roadless area resource center and database.

For further information, write The Mountaineers Foundation, P. O. Box 9464, Queen Anne Station, Seattle, WA 98109.